AUTOMOTIVE FUEL ECONOMY

HOW FAR SHOULD WE GO?

Committee on Fuel Economy of Automobiles and Light Trucks
Energy Engineering Board
Commission on Engineering and Technical Systems

National Research Council

NATIONAL ACADEMY PRESS
Washington, D.C. 1992

National Academy Press • 2101 Constitution Avenue, N.W. • Washington, D.C. 20418

NOTICE: The project that is the subject of this report was approved by the Governing Board of the National Research Council, whose members are drawn from the councils of the National Academy of Sciences, the National Academy of Engineering, and the Institute of Medicine. The members of the committee responsible for the report were chosen for their special competencies and with regard for appropriate balance.

This report has been reviewed by a group other than the authors according to procedures approved by a Report Review Committee consisting of members of the National Academy of Sciences, the National Academy of Engineering, and the Institute of Medicine.

The National Academy of Sciences is a private, nonprofit, self-perpetuating society of distinguished scholars engaged in scientific and engineering research, dedicated to the furtherance of science and technology and to their use for the general welfare. Upon the authority of the charter granted to it by the Congress in 1863, the Academy has a mandate that requires it to advise the federal government on scientific and technical matters. Dr. Frank Press is president of the National Academy of Sciences.

The National Academy of Engineering was established in 1964, under the charter of the National Academy of Sciences, as a parallel organization of outstanding engineers. It is autonomous in its administration and in the selection of its members, sharing with the National Academy of Sciences the responsibility for advising the federal government. The National Academy of Engineering also sponsors engineering programs aimed at meeting national needs, encourages education and research, and recognizes the superior achievements of engineers. Dr. Robert M. White is president of the National Academy of Engineering.

The Institute of Medicine was established in 1970 by the National Academy of Sciences to secure the services of eminent members of appropriate professions in the examination of policy matters pertaining to the health of the public. The Institute acts under the responsibility given to the National Academy of Sciences by its congressional charter to be an advisor to the federal government and, upon its own initiative, to identify issues of medical care, research, and education. Dr. Kenneth Shine is the president and chairman of the Institute of Medicine.

The National Research Council was organized by the National Academy of Sciences in 1916 to associate the broad community of science and technology with the Academy's purposes of furthering knowledge and advising the federal government. Functioning in accordance with general policies determined by the Academy, the Council has become the principal operating agency of both the National Academy of Sciences and the National Academy of Engineering in providing services to the government, the public, and the scientific and engineering communities. The Council is administered jointly by both Academies and the Institute of Medicine. Dr. Frank Press and Dr. Robert M. White are chairman and vice chairman, respectively, of the National Research Council.

This is a report of work supported by Grant No. DTNH22-91-Z-06014 from the National Highway Traffic Safety Administration and the Federal Highway Administration of the U.S. Department of Transportation to the National Research Council. The work was also supported by the Casey Fund of the National Research Council.

ISBN 0-309-04530-4

Printed in the United States of America

First Printing, April 1992
Second Printing, August 1992
Third Printing, November 1992

COMMITTEE ON FUEL ECONOMY OF AUTOMOBILES AND LIGHT TRUCKS

Chairman

RICHARD A. MESERVE, Partner, Covington & Burling, Washington, D.C.

Members

GARY L. CASEY, former director, Advanced Technology, Allied-Signal, Inc., Troy, Michigan
W. ROBERT EPPERLY, President, Epperly Associates, Inc., New Canaan, Connecticut
THEODORE H. GEBALLE, Professor of Applied Physics, Department of Applied Physics, Stanford University, Stanford, California
DAVID L. GREENE, Senior Research Staff, Oak Ridge National Laboratory, Oak Ridge, Tennessee
JOHN H. JOHNSON, Presidential Professor and Chairman, Department of Mechanical Engineering - Engineering Mechanics, Michigan Technological University, Houghton, Michigan
MARYANN N. KELLER, Managing Director, Furman Selz Incorporated, New York, New York
CHARLES D. KOLSTAD, Associate Professor, Institute for Environmental Studies and Department of Economics, University of Illinois, Urbana, Illinois
LEROY H. LINDGREN, Vice President, Rath & Strong, Inc., Lexington, Massachusetts
G. MURRAY MACKAY, Head, Accident Research Unit, Automotive Engineering Center, University of Birmingham, Birmingham, England
M. EUGENE MERCHANT, Senior Consultant, Institute of Advanced Manufacturing Sciences, Cincinnati, Ohio
DAVID L. MORRISON, Technical Director, Energy, Resource and Environmental Systems Division, The MITRE Corporation, McLean, Virginia
PHILLIP S. MYERS, Emeritus Distinguished Professor of Mechanical Engineering, University of Wisconsin, Madison, Wisconsin
DANIEL ROOS, Director, Center for Technology Policy and Industrial Development, Massachusetts Institute of Technology, Cambridge, Massachusetts
PATRICIA F. WALLER, Director, University of Michigan Transportation Research Institute, University of Michigan, Ann Arbor, Michigan
JOSEPH D. WALTER, Director of Central Research, Bridgestone-Firestone, Inc., Akron, Ohio

National Research Council Staff

Energy Engineering Board

MAHADEVAN (DEV) MANI, Study Director
CHRISTOPHER T. HILL, Executive Director,
The Manufacturing Forum (on loan)
GEORGE LALOS, Project Manager
JAMES ZUCCHETTO, Senior Program Officer
JUDITH AMRI, Study Administrative Assistant
SUSANNA E. CLARENDON, Administrative Assistant
STEPHEN CARRUTH, Study Assistant (January-June, 1991)
THERESA FISHER, Study Assistant
PENELOPE J. GIBBS, Administrative Assistant,
The Manufacturing Forum (on loan)
PHILOMINA MAMMEN, Study Assistant
NANCY WHITNEY, Study Assistant (January-September, 1991)
JEAN M. SHIRHALL, Consulting Editor

Commission on Engineering and Technical Systems

ARCHIE WOOD, Executive Director

ENERGY ENGINEERING BOARD

Liaison Members from the Commission on Engineering and Technical Systems

KENT F. HANSEN, Professor of Nuclear Engineering, Energy Laboratory, Massachusetts Institute of Technology, Cambridge, Massachusetts
JOHN B. WACHTMAN, Sosman Professor of Ceramics, Rutgers University, Piscataway, New Jersey

Energy Engineering Board Staff

MAHADEVAN (DEV) MANI, Director
KAMAL ARAJ, Senior Program Officer
GEORGE LALOS, Senior Program Officer
JAMES ZUCCHETTO, Senior Program Officer
JUDITH AMRI, Administrative and Financial Assistant
SUSANNA CLARENDON, Administrative Assistant
STEPHEN CARRUTH, Research Assistant (January-June, 1991)
THERESA FISHER, Senior Project Assistant
PHILOMINA MAMMEN, Senior Project Assistant
NANCY WHITNEY, Senior Project Assistant (January-September, 1991)
NORMAN HALLER, Consultant

Commission on Engineering and Technical Systems

ARCHIE WOOD, Executive Director
MARLENE BEAUDIN, Associate Executive Director

PREFACE

Following upon the oil embargo imposed by the Organization of Petroleum Exporting Countries (OPEC) in 1973, the U.S. Congress took action to reduce the dependence of the United States on imported petroleum. Because the transportation sector, including in particular the automotive subsector, is one of the main consumers of petroleum, Congress in the Energy Policy and Conservation Act of 1975 required that the automotive manufacturers improve the fuel economy of automobiles and light trucks that are sold in the United States. Congress set a requirement that the corporate average fuel economy (CAFE) of the new-car fleet of each manufacturer achieve 27.5 miles per gallon (mpg) in model year 1985 and thereafter. Congress did not, however, explicitly require continuing improvement of fuel economy after the target established in the act was accomplished.

The Congress and the Executive Branch are now reexamining issues of energy policy that have largely been ignored over the past decade. Because the dependence of the United States on foreign petroleum has grown despite advances in automotive fuel economy, Congress is now considering whether to require further improvements in fuel economy over the coming years. Indeed, the recent Persian Gulf war has shown the fragility of important sources of world petroleum supply and has perhaps reinforced the importance of reducing U.S. vulnerability to supply interruption. Moreover, the growing concern over global warming now provides an impetus for improving efficiency in the use of petroleum in view of the fact that automobiles and light trucks are the source of as much as 15 percent of the greenhouse gases emitted from anthropogenic sources in the United States.

In light of this renewed interest in automotive fuel economy, the U.S. Department of Transportation's National Highway Traffic Safety Administration and Federal Highway Administration requested the National Research Council (NRC) to undertake a study of the potential and prospects for improving the fuel economy of new light-duty vehicles. The NRC appointed the Committee on Fuel Economy of Automobiles and Light Trucks to conduct the study. This report is the result of the committee's deliberations in Phase I of the study.

The committee was asked to provide estimates by vehicle size class of the fuel economy that could "practically" be achieved in new automobiles and light trucks produced for the United States in the next decade. In fulfilling this task, the committee was asked, among other points, to examine the technologies that could contribute to improved fuel economy, to identify the barriers to their introduction, and to consider their costs and benefits. The evaluation was to include consideration of the effects of current environmental and safety requirements and the consequences of potential new requirements on the consumer and the automotive industry. (The full text of the committee's charge is set out in Appendix A; a description of the committee's membership is set out in Appendix G.)

The committee held its first meeting in Washington, D.C., in May 1991. This was followed by a five-day workshop at the Beckman Center in Irvine, California, in July, at which comprehensive presentations were made to the committee on a variety of relevant topics. Following the Irvine meeting, the committee met monthly through December 1991 to hold its deliberations and write this report. In addition, subgroups on technology, safety, emissions and environment, economic impacts, and standards/regulations were formed; the subgroups collectively held eight additional meetings during the course of the study. A chronology of the committee's work and a list of the many individuals who made presentations to the committee and its various subgroups are provided in Appendix F.

A few points about the committee's conclusions warrant mention at the outset. First, it is important to emphasize the caution with which the committee's estimates should be viewed. An examination of the accuracy of previous estimates of future automotive fuel economy provides a humbling demonstration that even knowledgeable analysts have often missed the mark by wide margins in making such predictions. For example, in the late 1970s, various observers estimated that the fuel economy of the new passenger-car fleet in 1990 would range from about 32 to 40 mpg, whereas a 27.8 mpg fleet average was in fact achieved. Many of the important variables in assessing such matters--for example, the price of gasoline--have not followed the expected path, with the consequence that related estimates, such as automotive fuel economy, have proven unreliable. Moreover, even assessments of the technologies that would be significant have proven erroneous; analysts in the 1950s predicted significant market shares by now for gas turbines, those of the 1960s predicted diesels would be in widespread automotive use, those of the 1970s emphasized rotary engines, and those of the 1980s pointed to turbochargers. The committee does not claim that its crystal ball is any less cloudy than those used by previous analysts; unforeseen events could cause its evaluations to be significantly in error.

Second, and wholly apart from the difficulties of predicting the future, difficulties are presented by reason of the many factors that must be considered in assessing the fuel economy that is practically achievable. As will be explained more fully in this report, the judgment as to what fuel economy gains are practical involves the assessment of a complex manifold of considerations: the perceived severity of the environmental impacts of energy use; the political and economic implications of growing dependence on imported oil; the availability and costs of fuel efficiency technologies; the price of gasoline; the evolution, impacts, and consequences of environmental and safety considerations; the time and capital required for the automotive industry to change its production facilities; the attitudes and interests of consumers; the financial capability of the domestic industry to accommodate new requirements; the consequences of new requirements on automotive workers and suppliers; the differential competitive impacts on foreign and domestic manufacturers;

and many other matters. Unfortunately, most of these considerations do not lend themselves to precise or reliable evaluation.

Third, even if the committee had the capacity to predict the future and to estimate the various relevant factors, the determination of what is practically achievable requires a significant measure of judgment on which reasonable people may disagree. The central theme of this report is that any strategy to increase fuel economy will impose costs and benefits--as will a policy of not encouraging improved fuel economy. Although the committee attempted to estimate the various costs and benefits at least qualitatively, ultimately the policymaker must decide how much enhanced fuel economy is worth. The determination of the appropriate balance of costs and benefits requires a measure and trade-off of values that extend beyond the special competence of this committee.

In light of these facts, the committee aimed in this report to provide information to guide and assist policymakers. In the final analysis, it hopes only that its efforts will be viewed as having defined and illuminated the issues, rather than having finally resolved them.

A few aspects of the study also warrant mention at the outset. Although the committee was charged with examining the fuel economy that is practically achievable over the next decade, it expanded the scope of its evaluation to 2006. As will be explained, over the next four years there is in fact little opportunity to increase fuel economy beyond the levels already planned by the manufacturers. The decade of opportunity for improved fuel economy starts in 1996.

The committee also examined only conventionally powered automobiles and light trucks, that is, those using gasoline and diesel fuel. It did not consider electric or gasoline-electric hybrid vehicles or vehicles powered by alternative fuels (such as methanol, ethanol, or natural gas). Although such vehicles might make a contribution, their evaluation would have expanded the scope of a study that already was daunting in its breadth. Such vehicles might be included in future work.

Although its charge did not require the committee to analyze the current structure and form of fuel economy standards or to address possible alternative approaches to improving fuel economy, the committee concluded that commentary on such matters was appropriate. The means by which enhanced fuel economy is achieved are perhaps as important as the matters the committee was asked to address. Thus, in the committee's opinion, a discussion of such matters is essential to provide perspective on the findings and recommendations requested by its charge. The committee's commentary on policies for improving fuel economy is set out in the concluding chapter. The commentary is by no means definitive because it is limited to an elucidation of the larger context in which fuel economy regulation ought to be considered by policymakers.

It only remains to be said that the committee received very substantial assistance in its efforts from the automotive manufacturers, from industry more generally, from environmental and safety groups, from the academic community, and from agencies of the U.S. government. The efforts of the many contributors were substantial and lightened the committee's load considerably. Moreover, the committee was assisted by an able, energetic, and enthusiastic staff. The committee appreciates their help.

RICHARD A. MESERVE, *Chair*
Committee on Fuel Economy of Automobiles and Light Trucks

TABLE OF CONTENTS

APPENDIXES

LIST OF TABLES

LIST OF FIGURES

AUTOMOTIVE FUEL ECONOMY

HOW FAR SHOULD WE GO?

EXECUTIVE SUMMARY

The U.S. Department of Transportation's National Highway Traffic Safety Administration (NHTSA) and Federal Highway Administration requested the National Research Council (NRC) to undertake a study of the potential and prospects for improving the fuel economy of new light-duty vehicles produced for the U.S. market. This report presents the results of the study conducted by the NRC's Committee on Fuel Economy of Automobiles and Light Trucks.

The charge to the committee was to estimate "practically achievable" fuel economy levels in various size classes of new passenger cars and light trucks using gasoline and diesel fuel. Any such determination of practically achievable fuel economy levels, however, necessarily involves balancing an array of societal benefits and costs, while keeping in mind where the costs and benefits fall. Such judgments must include a complex manifold of considerations, such as the financial costs to consumers and manufacturers, the impact on employment and competitiveness, the trade-offs of fuel economy with occupant safety and environmental goals, and the benefits to our national and economic security of reduced dependence on petroleum. *In the committee's view, the determination of the practically achievable levels of fuel economy is appropriately the domain of the political process, not this committee.*

TECHNICALLY ACHIEVABLE FUEL ECONOMY LEVELS

In order to illuminate the issues for decision makers, the committee has thus approached its charge by first seeking to estimate future "technically achievable" fuel economy levels by vehicle size class for automobiles and light trucks. As explained below, these estimates provide guidance, subject to certain assumptions, on the fuel economy that could be achieved using current technology. The committee also sets out its judgment of the range of retail price increases of new vehicles attributable strictly to such fuel economy improvements. (These estimates do not include the price increases to meet occupant safety and emissions control requirements.) The committee's time horizon extended from model year (MY) 1996 to MY 2006, the end point of the decade considered in the study.[1]

The term "technically achievable" is circumscribed by three assumptions: (1) that all new vehicles in the future fleets will have to comply with Tier I emissions standards under the Clean Air Act amendments of 1990 and existing and pending standards for occupant safety, (2) that current vehicle characteristics (such as interior volume or acceleration performance) valued by consumers will remain essentially unchanged, and (3) technologies considered for improving fuel economy are those that are currently used in mass-produced vehicles somewhere in the world and that pay for themselves at gasoline prices of $5 to $10 per gallon or less. *Aside from the limits imposed by these assumptions, no cost-benefit considerations involving new vehicle affordability, sales, employment in the automotive industry, competitiveness, or safety impacts entered into the determination of the "technically achievable" fuel economy levels.* In the committee's view all such matters should be considered, however, as policymakers establish the practically achievable fuel economy levels.

In light of the foregoing, the "technically achievable" fuel economy levels should not be taken as the technological limit of what is possible with the current state of the art; nor should the committee's estimates of what is technically achievable be taken as its recommendations on future fuel economy standards. Rather, they should be viewed by policymakers as one of the many inputs necessary for determining what would be practically achievable in the coming decade.

The committee believes that the practically achievable levels--the levels of fuel economy for each size class that achieve an appropriate balance of a broad array of costs and benefits--are likely to be found in the regions between the levels that would be achieved without any governmental intervention and the "technically achievable" levels. It remains for policymakers to determine the form of any future regulations and the levels of fuel economy--the "practically achievable" levels--that in their judgment provide the appropriate balance of costs and benefits to consumers, manufacturers, and the nation as a whole.

The committee analyzed, qualitatively and quantitatively, available technological and cost information in arriving at its estimates of the fuel economy improvements that it judged, on average, to be technically achievable in new vehicles. The estimates represent the collective professional judgment of the committee in light of available evidence, including presentations to the committee on market-penetration potential, costs, and effectiveness of technologies to improve fuel economy. They also reflect the committee's consideration of past trends in fuel economy improvements in the United States and fuel economy levels being achieved by today's "best-in-class" vehicles. Table

---

[1]The committee considered that the decade of opportunity for increasing fuel economy begins with MY 1996. Given that the product plans of the automobile manufacturers through MY 1995 are already largely set, little can be done to improve automotive fuel economy beyond what is already planned--at reasonable cost--before MY 1996.

ES-1 summarizes, by size class, the committee's range of estimates for technically achievable fuel economy in MY 2006 and the costs, expressed in average incremental vehicle retail price equivalents (RPEs), for the fuel economy technologies associated with those estimates. The ranges embody measures of the committee's confidence of what might transpire in the marketplace, given the bounds of this study and its underlying assumptions.

The committee also considered available information on occupant safety and emissions control. Compliance with safety regulations is expected to add about $300 to the average price of new passenger cars and about $500 to the price of light trucks, in 1990 dollars. Cost estimates for complying with Tier I emission controls vary widely--from a few hundred dollars to $1,600 per vehicle--a reflection of the considerable range of uncertainty about them. The potential costs of Tier II standards are even less well understood, but they are thought to be substantial. *The retail price equivalents in Table ES-1 do not include the incremental costs of improved safety and emissions control.*

The realization of the technically achievable fuel economy levels shown in Table ES-1 will be affected by a number of factors. These include the rate and success of technology development, the competitive strategies of various manufacturers, fuel prices and availability, performance of the U.S. economy, trends in consumer tastes, and the form of future fuel economy standards themselves, as well as the effects of safety and emissions regulations.

The sections that follow highlight the committee's conclusions with respect to factors that must be considered by policymakers in determining practically achievable fuel economy levels for light-duty vehicles.

PROVEN AND EMERGING TECHNOLOGIES FOR IMPROVING FUEL ECONOMY

Modern automobiles and light trucks are complex, multipurpose vehicles used to move passengers and goods comfortably and safely. They have been designed to meet a complex set of consumer and regulatory requirements, and they are technologically mature products. While super-efficient cars and concept vehicles have been demonstrated, they have not met many requirements of the market. Thus, such vehicles in and of themselves are unreliable guides for estimating future fuel economy potential.

Many proven technologies are available to improve fuel economy, each of which can make small, but important, contributions. Improvements are most likely to occur as a result of many changes that affect all the principal determinants of fuel economy--namely, engine and drivetrain efficiency, vehicle weight, aerodynamic drag, rolling resistance of tires, and the efficiency of accessories.

A number of emerging technologies hold the promise of better fuel economy--for example, lean-burn gasoline engines, two-cycle engines, and advanced, environmentally acceptable diesels. However, in part because of stringent new and

TABLE ES-1 "Technically Achievable" Fuel Economy for MY 2006 Vehicles

Vehicle Size Class	Ranges of "Technically Achievable" Fuel Economy in MY 2006[a] (mpg)		Incremental Retail Price Equivalent for Improved Fuel Economy in MY 2006[b] (1990 Dollars)	
	Higher Confidence	Lower Confidence	At Higher Confidence Fuel Economy	At Lower Confidence Fuel Economy
Passenger Cars				
Subcompact	39	44	500-1,250	1,000-2,500
Compact	34	38	500-1,250	1,000-2,500
Midsize	32	35	500-1,250	1,000-2,500
Large	30	33	500-1,250	1,000-2,500
Light Trucks				
Small pickup	29	32	500-1,000	1,000-2,000
Small van	28	30	500-1,250	1,000-2,500
Small utility	26	29	500-1,250	1,250-2,500
Large pickup	23	25	750-1,750	1,500-2,750

a/ The term "technically achievable" is circumscribed by the following assumptions made by the committee. The estimates result from consideration of technologies currently used in vehicles mass produced somewhere in the world and that pay for themselves at gasoline prices of $5 to $10 per gallon or less (1990 dollars). The estimates assume compliance with applicable known safety standards and Tier I emissions requirements of the Clean Air Act amendments of 1990. Compliance with Tier II and California's emissions standards has not been taken into account. The estimates also assume that MY 2006 vehicles will have the acceleration performance of, and meet customer requirements for functionality equivalent to, 1990 models. The estimates take into account past trends in vehicle fuel economy improvements and evidence from "best-in-class" fuel economy experience. The term "technically achievable" should not be taken to mean the technological limit of what is possible with the current state of the art; nor should the committee's estimates of what is technically achievable be taken as its recommendations as to what future fuel economy levels should be.

Aside from the limits imposed by the foregoing assumptions, no cost-benefit considerations entered into the determination of the technically achievable fuel economy levels. Specifically, the estimates do not take into account other factors that should be considered by policymakers in determining any future fuel economy regulations, including impacts on the competitiveness of automotive and related industries, sales and employment effects, petroleum import dependence, effects on nonregulated emissions (e.g., the greenhouse gas, carbon dioxide), and the development and adoption of unanticipated technology.

As a point of reference, the Environmental Protection Agency's (EPA's) composite average fuel economy for MY 1990 passenger cars and light trucks, by size class, was as follows: passenger cars--subcompact, 31.4 mpg; compact, 29.4; midsize, 26.1; large, 23.5; light trucks--small pickup, 25.7; small van, 22.8; small utility, 21.3; large pickup, 19.1 (Heavenrich et al., 1991).

b/ The retail price equivalents are estimates only of the incremental first cost to consumers of improved fuel economy. They do not include incremental costs associated with mandated improvements to occupant safety, which, on average for new passenger cars and light trucks, are expected to be $300 and $500, respectively in 1990 dollars; nor do they include incremental costs of controls to comply with Tier I emissions requirements, which are expected to range from a few hundred dollars to $1,600 per vehicle.

proposed emissions standards (especially for oxides of nitrogen, NO_x), it is impossible at this stage to estimate with any accuracy the probability of their success. Nonetheless, it seems reasonably certain that one or more emerging technologies (including some not foreseen in this study) will begin to make significant contributions to fuel economy by MY 2006.[2]

The committee had available several recent analyses of the potential for improving automotive fuel economy in the future. However, even estimates of the fuel economy contribution of technologies already in mass production vary substantially from source to source. Further, information on the costs of those technologies and their rate and scope of application is very limited; and, where such information is available, the estimates also vary widely.

There is little data collection or sustained analysis by cognizant governmental agencies on the costs and performance of fuel economy technologies and on developments in the automotive industry in general. Most of the studies on the potential and prospects for improving fuel economy that were performed outside the industry and virtually all the debates in Congress on the subject have drawn on the work of one firm operating under contract to the Department of Energy. It is inescapable that Congress and the governmental agencies are attempting to regulate an industry of tremendous importance to the U.S. economy in the absence of sufficient information from neutral sources on which to base such regulation. The committee recommends that the Department of Transportation, the National Highway Traffic Safety Administration, and the Federal Highway Administration, in concert with other federal agencies, such as the Department of Energy and the Department of Commerce, reestablish a robust program of data collection and analysis to support the formulation of national policy.

SAFETY IMPLICATIONS

Safety, as measured by fatalities in the United States per hundred million vehicle miles driven, has been steadily improving for decades--falling from about 16 fatalities per hundred million miles in 1930 to about 2 in 1990. It is likely that this general trend will continue, independently of other measures. Nonetheless, safety and fuel economy are linked because one of the most direct methods manufacturers can use to improve fuel economy is to reduce vehicle size and weight. In certain crash types--for example, a collision with a nonrigid fixed object--vehicle weight is a major determinant of the forces experienced by vehicle occupants. Any major reduction in vehicle weight carries the potential for reduced safety in such collisions. Such changes stir controversy because of the differences in the occupant risk between driving in a smaller (higher fuel economy) car and driving in a larger (lower fuel economy) car.

In addition, several studies of single-vehicle accidents show a higher risk of occupant injury and death in small vehicles. This is presumably due in part to their

[2]The committee did not evaluate hybrid or electric vehicles, whose role as yet remains largely undefined.

greater propensity to roll over. It is the committee's understanding that this risk can be reduced through improved vehicle design and that NHTSA has promulgated an advance notice of proposed rulemaking on the matter. There could be a penalty for improved fuel economy in this regard, but it should be small.

In two-car collisions the evidence shows that, from the standpoint of the individual, a large car affords greater protection (or lower risk of a fatal injury) to its occupants than a small car. However, it is still a matter of some debate as to whether the higher level of safety accorded by the large car comes predominantly from its greater weight, its larger size (external dimensions), or both attributes. The available data do not allow a definitive resolution of the issue because size and weight reductions are highly correlated.[3]

While it seems clear that the occupants of a large car are safer than those of a small car with equal safety provisions, it is not so clear whether and how much societal risks are materially changed by substituting small cars for large. The perspective of the society at large differs from that of the individual because a large car, while providing an increment of safety to its occupants, imposes an increment of risk (compared with a small car) on others. Little research has been done on such effects, but a hypothetical analysis of two-car collisions by the committee suggests that, in principle, downsizing could increase, decrease, or leave unchanged total deaths and injuries in two-car collisions, depending on the changed size distribution of cars in the fleet. This presents only a partial picture, however. Fatalities from two-car collisions represent only a portion (about 11 percent) of total fatalities; they do not include other types of crashes--single vehicle into roadside obstacles, passenger car-truck collisions, and collisions of cars and light trucks with pedestrians and cyclists. It is the committee's view that the net safety consequences of downsizing and downweighting merit comprehensive analysis by NHTSA. Such analysis should take into account the entire population of vehicles on the road, as well as the incidence of accidents and fatalities involving pedestrians and cyclists.

Although the data and analyses are not definitive, the committee believes that there is likely to be a safety cost if downweighting is used to improve fuel economy (all else being equal). The committee believes, however, that the safety impacts of fuel economy improvements at the levels shown in Table ES-1 may be small, since they entail no more than a 10 percent reduction in weight, on average, in any size class and since improved design and safety technology present the opportunity to reduce the effects of weight reduction. As noted earlier, the available information was insufficient to make specific estimates of those impacts.

Light trucks have special safety problems, and because they constitute almost one-third of the new vehicle fleet, they warrant careful scrutiny. Moreover, light trucks are particularly aggressive to passenger cars in car-light truck collisions. Some measures

---

[3] The issue has significance because weight reduction is important for fuel economy improvement and could be achieved without size reduction through the use of lighter (and more expensive) materials.

to reduce such aggressivity--for example, lowering bumpers--should not affect fuel economy.

Currently scheduled and anticipated modifications to improve vehicle safety include passenger-side airbags, improved side-impact protection, and antilock braking systems. Improved design and the incorporation of new technology can also enhance crash-avoidance potential, as well as provide occupant protection in a collision. Most such changes will increase vehicle weight and thereby decrease fuel economy.

Safety is an important consideration in fuel economy deliberations, but it must be considered in relation to other important societal values that are affected by improved fuel economy levels. Trade-offs of safety with other societal objectives are not unusual. Recent examples include increasing the national maximum speed limit on rural interstate highways from 55 to 65 miles per hour and permitting right-turn-on-red. Both measures were predicted to increase traffic fatalities and injuries and subsequent studies confirmed those predictions. Thus, concern for safety should not be allowed to paralyze the debate on the desirability of enhancing the fuel economy of the light-duty fleet.

ENVIRONMENTAL ISSUES

Fuel economy improvements will not directly affect vehicle emissions of hydrocarbons, carbon monoxide, and NO_x because the emissions standards (in grams per mile) are identical for every passenger car or light truck, as appropriate, regardless of fuel economy. Fuel economy improvements in new light-duty vehicles will reduce carbon dioxide emissions per mile because less fuel will be consumed per vehicle mile driven.

In its fuel economy estimates presented in Table ES-1, the committee assumed that the Tier I emissions requirements imposed by the Clean Air Act amendments of 1990 will be met. (The committee took account of the penalty in fuel economy arising from increased weight of new or improved emissions controls on vehicles.) The committee made no allowance for on-board vapor recovery or meeting the stricter Tier II or new California standards. In the committee's view, standards more stringent than the Tier I requirements imposed in the Clean Air Act amendments of 1990 would have adverse implications for improvements in automotive fuel economy.

Without a major technological breakthrough (e.g., a lean NO_x catalyst), there will be tight constraints on the application, particularly to larger vehicles, of promising current and emerging technologies for improving fuel economy, such as lean-burn and advanced diesel engines. Improved emissions controls are needed for widespread use of two-stroke engines. It appears possible that a lean NO_x catalytic system achieving roughly a 50 percent reduction in NO_x emissions from the engine and with the required durability will be developed. Such a system might make it possible to meet Tier II and California's low-emission vehicle (LEV) standards for the smallest cars and trucks, but the system is unlikely to achieve the control needed in the heaviest light-duty vehicles without an additional, substantial breakthrough in catalyst technology.

Compliance with the Tier II and California standard of 0.2 grams per mile of NO_x may not be feasible for the current mix of automobiles and light trucks and is likely to be most difficult to meet with larger vehicles. Diesel-powered vehicles are unlikely to meet the Tier II and certain California standards for NO_x and particulates. While the Tier II standard is yet to be adopted nationally, the California standards are being implemented and are being seriously considered by other states that collectively account for about half of U.S. new-car sales. This constitutes a major uncertainty for the manufacturers. In view of the difficulty of meeting these standards, other approaches, such as increasing the level of control over NO_x emissions from stationary sources, should be considered.

In addition, the importance of controlling pollutant emissions from all sources suggests that (1) service station controls of refueling emissions of hydrocarbons should be given further consideration by the Environmental Protection Agency (EPA); (2) further reduction of sulfur in gasoline should be considered by EPA as part of its reformulated gasoline program; and (3) surveillance and enforcement of emissions standards in the existing fleet may be an attractive alternative to increasingly stringent controls on new vehicles.

THE AUTOMOTIVE INDUSTRY

The automotive industry is cyclical and has endured many upturns and downturns in its history. It is currently in an unprecedented downturn, suffering losses of billions of dollars. Competitive pressures on the domestic manufacturers, particularly from the Japanese manufacturers, are likely to intensify in the future. Moreover, productive capacity worldwide and particularly in the United States exceeds prospective demand for new vehicles, which is projected to grow only modestly over the period considered in this study. For these reasons alone, the U.S. companies are undertaking a major restructuring that will close some plants and call for significant capital investments to modernize others. As a result, employment in the sector will be reduced.

Since the U.S. automotive industry directly and indirectly employs substantial numbers of American workers, such changes are expected to have major impacts on the U.S. economy. These impacts will occur whether or not the government decides to increase future fuel economy standards. They could be aggravated, however, if more stringent standards or standards of an inappropriate form lead to significant increases in vehicle purchase prices that lower sales, or if the standards contribute to a shift toward greater purchases of vehicles manufactured outside the United States.

The interactions of higher fuel economy, improved occupant safety, and lower emissions illustrate dramatically the need for coordinated action by policymakers. There are limits to what can be achieved, and there are trade-offs that must be made. In setting fuel economy, safety, emissions, and other standards that add to the cost of the automobile, Congress and the administration should consider the cumulative impacts. Because the standards are currently set by different agencies under differing statutory commands, little coordination of policy is achieved and inconsistent pressures

--such as those arising from environmental and fuel economy considerations--are allowed to persist. This is indeed unfortunate because the automotive industry accounts for significant employment and is the focus of much of the national debate on improvements to the economy, the environment, and the international competitiveness of U.S. industry.

The imposition of higher fuel economy standards that are extremely costly or that greatly distort normal product cycles would place an untenable financial burden on the industry. Within 10 to 15 years, all current models and most engines and drivetrains will undergo at least one major change, and the equipment used in their manufacture will be written off. If the timing of new fuel economy standards follows the industry's product-development schedule, some, but not all, of the financial risk of new standards would be reduced. New regulations on fuel economy should therefore not require premature retirement of plant and equipment such that the industry is unable to recover its sunk costs. The industry is in a poor position to confront significant new investment requirements.

The committee recognizes the demands on industry and has taken some of them into account. Specifically, the committee supports levels of fuel economy improvement consistent with the manufacturers' product plans through MY 1995. Moreover, in considering a time horizon to MY 2006, the committee has implicitly factored in sufficient lead time to allow manufacturers to integrate changes into their normal product and plant replacements.

THE CONSUMER

Under the current regime of low fuel prices, consumers have relatively limited interest in purchasing, or manufacturers in producing, cars and light trucks with high fuel economy. Moreover, if gains in fuel economy are obtained in new vehicles at the expense of other attributes that consumers value (e.g., performance, size, safety, accessories, and cargo space), consumers may tend to retain their current vehicles longer or find themselves with little choice but to purchase vehicles that do not satisfy their requirements.

Since the early 1980s, there has been a major shift in consumer demand toward light trucks, most of which are used in the same way as automobiles. Because light trucks have lower fuel economy, on average, than automobiles, the trend has adversely affected fuel economy goals. Moreover, there has been increasing consumer demand for options that negatively affect fuel economy.

Consumer trends, including an aging population, increased demands for safety and for higher performance, and increased purchases of light trucks, are all in opposition to achieving higher fleet fuel economy. In the face of such trends, it is risky for manufacturers to invest in the design and production of vehicles with high fuel economy because they may not be able to sell them profitably. If higher fuel economy standards are to be imposed on manufacturers, it seems prudent that they be coupled with incentives such as cash rebates, higher fuel prices, or both, for consumers to purchase more fuel-efficient vehicles.

POLICIES FOR IMPROVING FUEL ECONOMY

The existing system for setting fuel economy standards--the corporate average fuel economy (CAFE) system--has serious defects that warrant careful examination. Chief among these defects is the fact that the CAFE system is increasingly at odds with market signals, which serves to mute and diminish the system's effectiveness and to increase its costs.

The existing CAFE standards, which require the domestically produced and imported vehicles sold by each manufacturer to achieve a specified average fuel economy rating, can also have perverse competitive effects. The standards have their most severe impact on the full-line manufacturers--the manufacturers that include large cars among their offerings. Because large cars generally have lower fuel economy than small cars, full-line manufacturers must invest resources to increase their sales of small cars and/or invest in technology to increase the fuel economy of their large cars. Manufacturers of small cars, on the other hand, can more readily meet the standards and indeed, may even produce fleets with fuel economy sufficiently above the CAFE standards as to enable them to expand initially into the large-car market without applying the expensive technology required of the full-line manufacturers. The existing system thus does not present equivalent technical or financial challenges to all manufacturers: Full-line manufacturers must strain to comply, whereas small-car manufacturers can comply with comparative ease. As it happens, this characteristic of the CAFE system has operated to the benefit of the Japanese manufacturers and to the detriment of the domestic (and some European) manufacturers. The CAFE system thus enhanced the competitive position of those foreign manufacturers that now pose the greatest challenge to the domestic industry.

The percentage-improvement approach to CAFE regulation--an approach proposed in certain legislation pending in the Congress--has the perverse effect of requiring those manufacturers with the best fleet fuel economy in a base year to comply with CAFE requirements in the outlying years that are more stringent than those applied to manufacturers with either lesser base-year accomplishments or that specialize in larger cars. The selection of the base-year creates arbitrary advantages and disadvantages, depending on the happenstance of the product mix or technology that a manufacturer was using in the base year. The percentage-improvement approach could be unfair because the manufacturers facing the most stringent requirements may already have incorporated many of the available fuel efficiency technologies in the base year. Moreover, it would seem to limit competition in the large-car classes by impeding the ability of the Japanese manufacturers to increase their market share in those classes.

Other alternatives to the existing CAFE system are worthy of consideration. A revised system might establish fuel economy requirements that are tailored to an attribute of the car, such as size class or usable passenger volume (or passenger-carrying capacity). Such an approach has an advantage over the existing system in that it could be applied so as to ensure that each manufacturer is confronted with a roughly equal challenge to improve fuel economy, regardless of the differences in the sizes of cars that the manufacturer produces. However, such a system may or may not serve to ensure that a given overall fleet fuel economy level is achieved, and the approach could be susceptible to "gaming" to exploit the system.

A policy of increasing fuel prices warrants consideration as an alternative or as a supplement to vehicle-efficiency regulation. Increasing fuel prices would serve to internalize the costs associated with fuel usage and reduce vehicle miles traveled. It would also affect the use of all vehicles on the road immediately, not just the use and performance of new vehicles. And, it would provide a market signal to channel consumer behavior in a direction consistent with societal objectives. It is the committee's view that, properly considered, there are ways to increase fuel prices without necessarily increasing the total costs to consumers of owning and operating vehicles. However, the committee has had neither the time nor the capacity to evaluate fully the price increase for fuel that would induce fuel consumption that is equivalent to that associated with aggressive CAFE standards. It may well be the case that the necessary increases in fuel price are so great as to make sole reliance on a pricing strategy politically unacceptable because of the ripple effects on the economy of the adjustment to higher prices.

Similarly, a system of fees and rebates that is related to the fuel economy of vehicles might also be considered. (Consumers who purchase a vehicle with below-average fuel economy could be charged a fee, and consumers who purchase a vehicle with above-average fuel economy could be given a rebate; the system could be revenue neutral in that the fees could cover the rebates.) Such "feebates" would establish incentives to encourage the acquisition of fuel-efficient vehicles, and the system would provide continuing pressure for change.

If the basic CAFE system is retained, Congress should consider several modifications. In light of the increasing interest in and use of light trucks, the fuel economy requirements for such vehicles should be brought into conformance with those for automobiles. The domestic content and credit provisions should be reexamined. And, the law should be modified so that noncompliance with a CAFE limit is not unlawful conduct, and the penalty for noncompliance should be adjusted so that it better reflects the social cost of departure from the requirements. These changes would increase the flexibility for manufacturers to respond to the law in an economically efficient way.

The objective of reducing petroleum consumption can be achieved by a variety of means. In addition to the CAFE system (and its variants) and increased fuel prices, the available policy instruments include improving the transportation infrastructure, developing intelligent vehicle-highway systems, improving public transit, reducing speed limits, encouraging car-pooling, and so forth. All such policy instruments should be considered in developing an appropriate strategy for reducing petroleum consumption.

1

INTRODUCTION

In 1990, the U.S. transportation sector accounted for over 25 percent of total U.S. energy consumption (U.S. Department of Energy, 1991; Energy Information Administration [EIA], 1991c).[1] Light-duty vehicles (passenger cars and light trucks) account for about 55 percent of the energy demand from the transportation sector. Moreover, light-duty vehicles figure disproportionately in the use of petroleum, consuming an average of over 10 million barrels per day (MMbbl/day) out of the total U.S. consumption of about 17 MMbbl/day (EIA, 1991a). Roughly half of the U.S. demand for petroleum is met with imports, and that fraction is expected to grow.[2]

In 1975, in the wake of a petroleum supply interruption, Congress enacted the Energy Policy and Conservation Act.[3] That statute required that automotive manufacturers selling cars in the United States increase the corporate average fuel economy (CAFE) of their new-car fleet to 27.5 miles per gallon (mpg) in model year (MY) 1985 and thereafter, unless the requirement was relaxed by the Secretary of Transportation.[4] Although the standards required by the act were eventually achieved, U.S. dependence on petroleum, especially on foreign supplies, rose in the 1980s, after dropping significantly between the early 1970s and the early 1980s. U.S. consumption of petroleum grew by 4.8 percent between 1981 and 1990, and the portion met by net imports grew from 36 to 46 percent in the same period (EIA, 1990, 1991c). Moreover, despite the imposition of CAFE requirements, consumption of petroleum by light-

---

[1]In 1990, the United States used nearly 82 quadrillion British thermal units (quads) of energy, of which the transportation sector alone consumed over 22 quads.

[2]As a percentage of the total U.S. trade deficit, the energy component (largely petroleum) increased from 21 percent in 1986 to 50 percent in 1990 (EIA, 1991c). Total nct petroleum imports are expected to reach 13 MMbbl/day by 2010, out of total petroleum products supplied of 20 MMbbl/day (EIA, 1991a).

[3]15 U.S.C. §§ 2001 et seq. (1988). Various features of the act are discussed in Chapter 9.

[4]The term *model year* applies to vehicles produced over an annual production period as determined by the U.S. Environmental Protection Agency (EPA). See 40 C.F.R. § 600.002-85 (1991).

duty vehicles grew from 41 percent of total petroleum consumption to 43 percent in the period from 1975 to 1990.[5]

Although the imposition of CAFE requirements may have diminished the consumption of petroleum below the levels that would otherwise have occurred, the concerns that prompted Congress to enact legislation in 1975 requiring improvements in the efficiency of the automotive fleet remain. Indeed, the recent conflict in the Persian Gulf has reemphasized the fragility of the world's petroleum supply and the importance of reducing U.S. dependence on foreign supply. Moreover, efforts to reduce the consumption of carbon-based fuels now have increasing significance as a result of concerns about global warming. Renewed examination of the opportunities--and means--for reducing petroleum consumption is warranted.

FUEL ECONOMY TRENDS SINCE 1975

Between 1975 and 1991 the fuel economy of the average new car improved by roughly 76 percent, from 15.8 mpg to 27.8 mpg (see Figure 1-1).[6] The rate of improvement in fuel economy varied over the period, however. Significant gains were made during the early years, from about 1975 to 1981. More modest gains were made from 1981 to 1988, and declines occurred from 1988 to 1991.

The gain in fuel economy resulted from a number of changes to automobiles. Reductions in vehicle weight were accomplished without significant reductions in interior volume (Figures 1-2 and 1-3) by decreasing the exterior dimensions of cars, switching from rear-wheel to front-wheel drive, and using lightweight aluminum and plastic materials. Improvements in engine efficiency (including reductions in the number of cylinders), engine horsepower, and displacement were also made. The introduction of new catalytic systems allowed the design of engines for more efficient operation. In addition, better aerodynamic design, increases in drivetrain efficiency, and reductions in tire rolling resistance and friction losses in general all contributed to increased fuel economy.

The decline in fuel economy from 1988 to 1991 of about 0.9 percent per year from its peak of 28.6 mpg in 1988 mainly resulted from increases in the performance and weight of the new-car fleet during the period.

The changes in fleet fuel economy were achieved against a backdrop of other changes. As shown in Figure 1-4, the cost of gasoline increased in real terms through

---

[5]Gasoline supplied in 1975 and 1990 was 6.7 and 7.2 MMbbl/day, respectively. Petroleum products supplied in 1975 and 1990 were 16.3 and 17 MMbbl/day, respectively (EIA, 1991c).

[6]The level of fuel economy attributed to new vehicles in a given year is associated with a model year, not a calendar year.

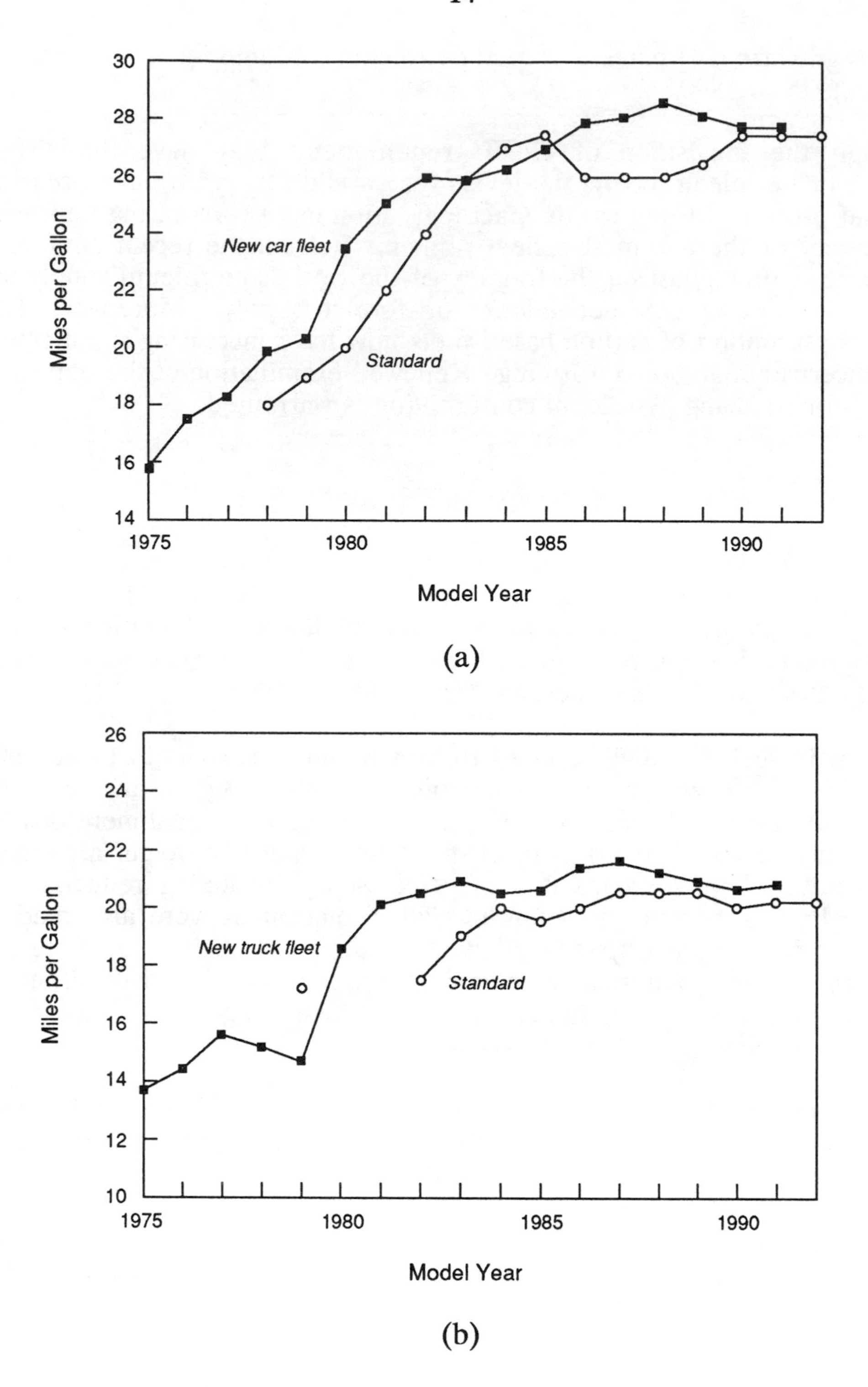

FIGURE 1-1 Trends in fuel economy for cars and light trucks.
NOTE: No standards were set for model years 1978, 1980, and 1981 for light trucks.

SOURCES: Based on Heavenrich et al. (1991) and Oak Ridge National Laboratory (ORNL, 1991).

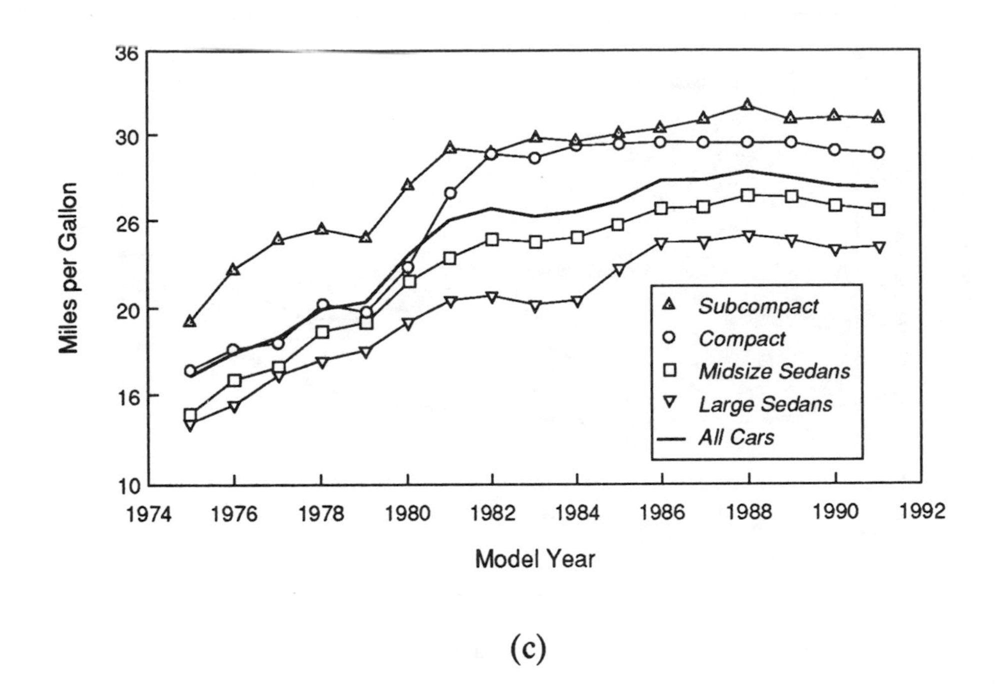

(c)

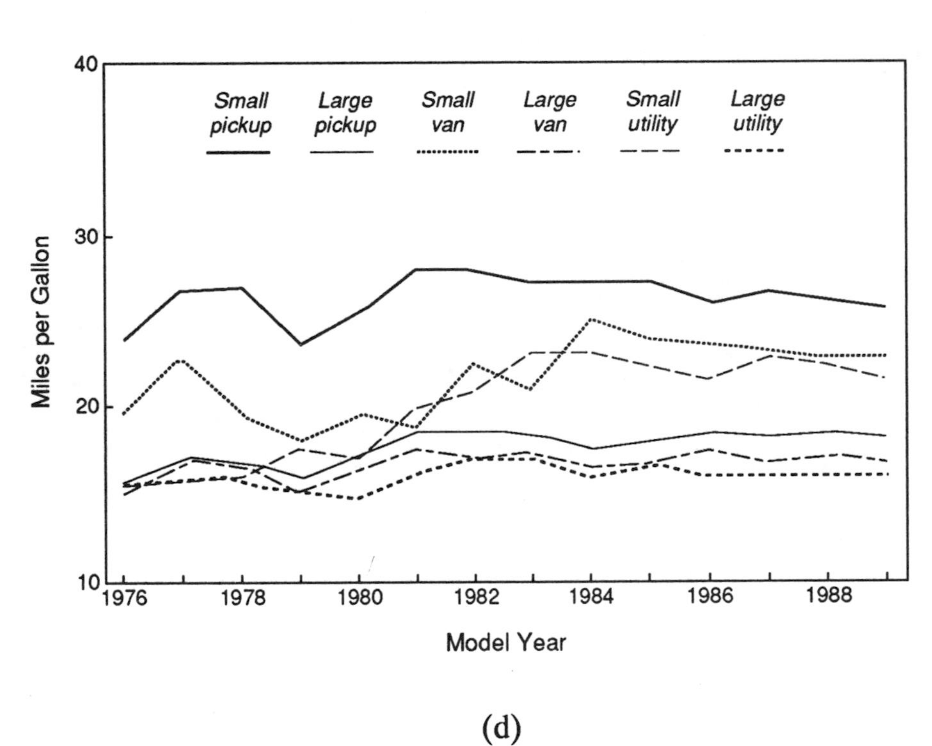

(d)

FIGURE 1-1 Trends in fuel economy for cars and light trucks **(continued)**.
NOTE: No standards were set for model years 1978, 1980, and 1981 for light trucks.

SOURCES: Based on Heavenrich et al. (1991) and ORNL (1991).

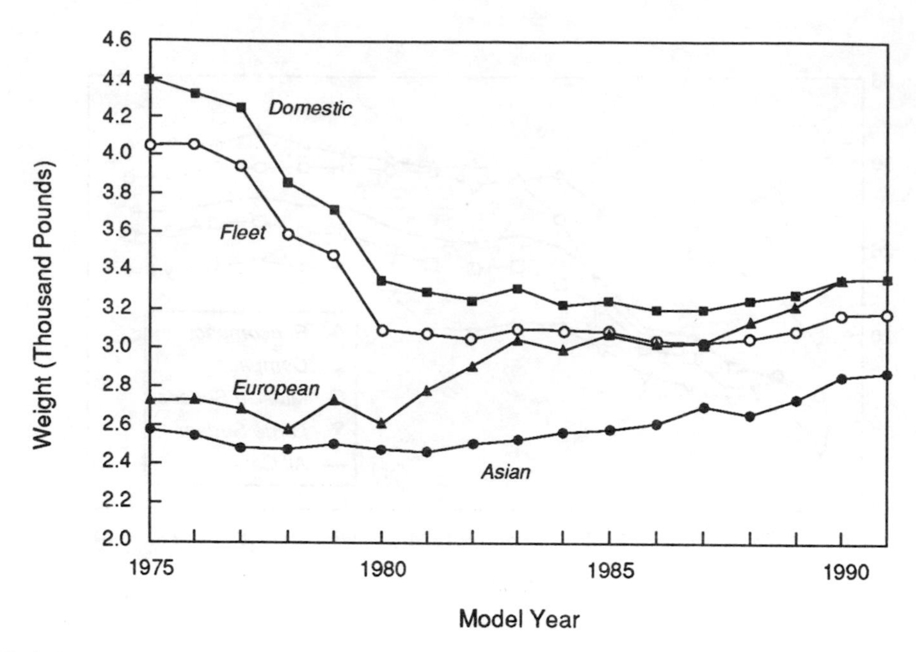

FIGURE 1-2 Trends in average weight for passenger cars sold in the United States, 1975-1991.

SOURCE: Based on Heavenrich et al. (1991).

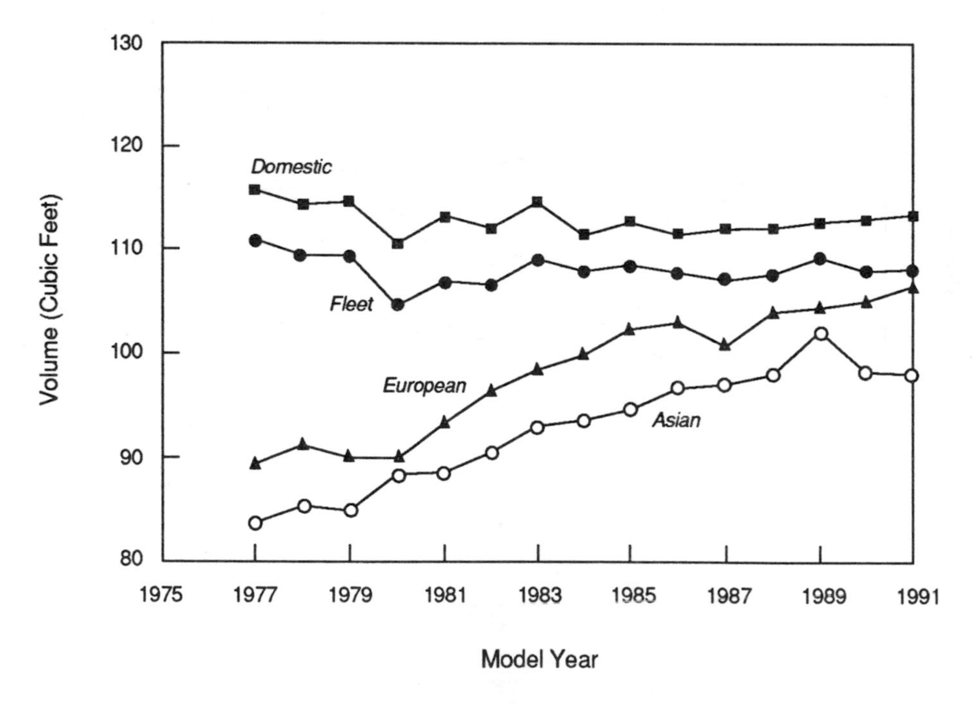

FIGURE 1-3 Trends in average interior volume for passenger cars sold in the United States, 1975-1991.

SOURCE: Based on Heavenrich et al. (1991).

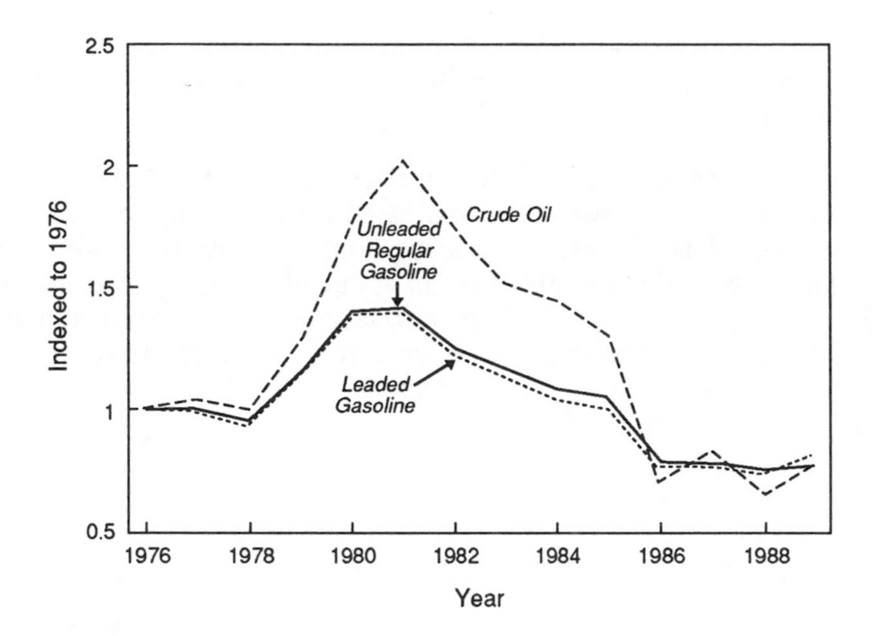

FIGURE 1-4 U.S. crude oil and gasoline prices, 1976-1989 (based on constant 1988 dollars).

SOURCE: ORNL (1991).

about 1981, but declined significantly thereafter. Thus, although gasoline prices in the late 1970s seem to have reinforced consumer interest in improved fuel economy,[7] the declining real price of gasoline since 1981 has discouraged further improvements. Indeed, because real fuel costs have declined and in any event have become an increasingly smaller component of the annual cost of vehicle ownership and operation (ORNL, 1991), fuel economy now has a low priority in the automobile purchase decisions of consumers.[8] In other words, efforts to improve fuel economy are now running against the market. Further, the perception of the late 1970s and early 1980s of the vulnerability of energy supplies has dissipated.

[7]Other factors were no doubt involved in the rapid improvement, such as the gasoline lines of the 1970s and the imposition of CAFE standards.

[8] Surveys by J.D. Power and others have found that fuel efficiency is given low priority by consumers in automobile purchase decisions. This information was presented to the committee at its workshop in Irvine, Calif., July 8-12, 1991 (see Appendix F); see also, for example, U.S. Department of Transportation (1991).

The improvements (and recent declines) in fleet fuel economy also occurred in a period of changing consumer interest in the various size classes of cars.[9] Subcompacts were of particular interest to consumers in the early 1980s, but consumer interest in such cars has diminished as interest in compact cars has grown (Figure 1-5). Consumer interest in light trucks since the early 1980s has reduced the average fuel economy of the new light-duty fleet (see below).

Since 1975, average engine size (displacement) has decreased, especially initially (Figure 1-6). Although engine power also decreased initially (Figure 1-7), it has since increased as technology has allowed significant enhancement of power per unit of displacement (Figure 1-8). The combination of reduced weight and increased power per unit of displacement resulted in improved acceleration performance (shorter acceleration time from 0 to 60 mph), especially during the latter portion of the time period (Figure 1-9). This improved performance reduced the fuel economy that otherwise could have been achieved.[10]

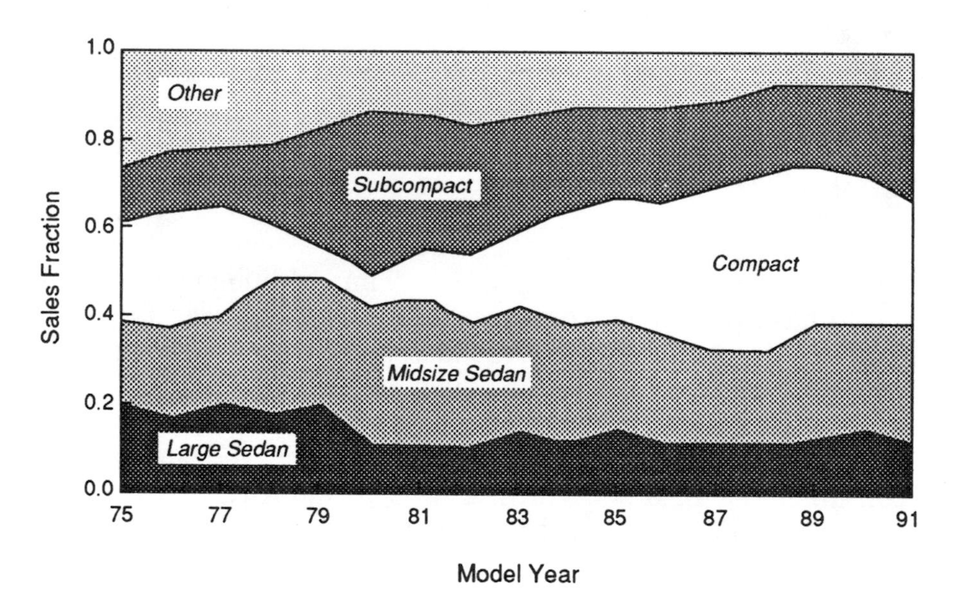

FIGURE 1-5 Sales fraction by car class.

SOURCE: Heavenrich et al. (1991).

---

[9]*Size class* refers to a system of classification of vehicles applied by the EPA (40 C.F.R. § 600.315-82 [1991]). The classes are defined on the basis of interior volume; most cars fall into the classes defined as subcompact, compact, midsize, and large. The principal light-truck classes that are the focus of this study are small pickup, large pickup, small van, and small utility.

[10]The average horsepower-to-weight ratio--another measure of performance capability--is greater now than at any time since 1975.

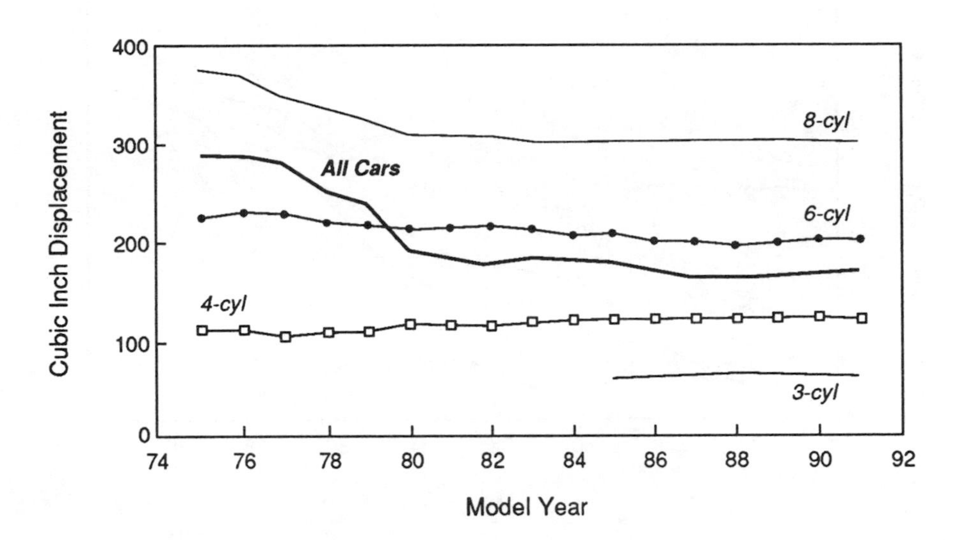

FIGURE 1-6 Average engine size for passenger cars, 1975-1991.

SOURCE: Heavenrich et al. (1991).

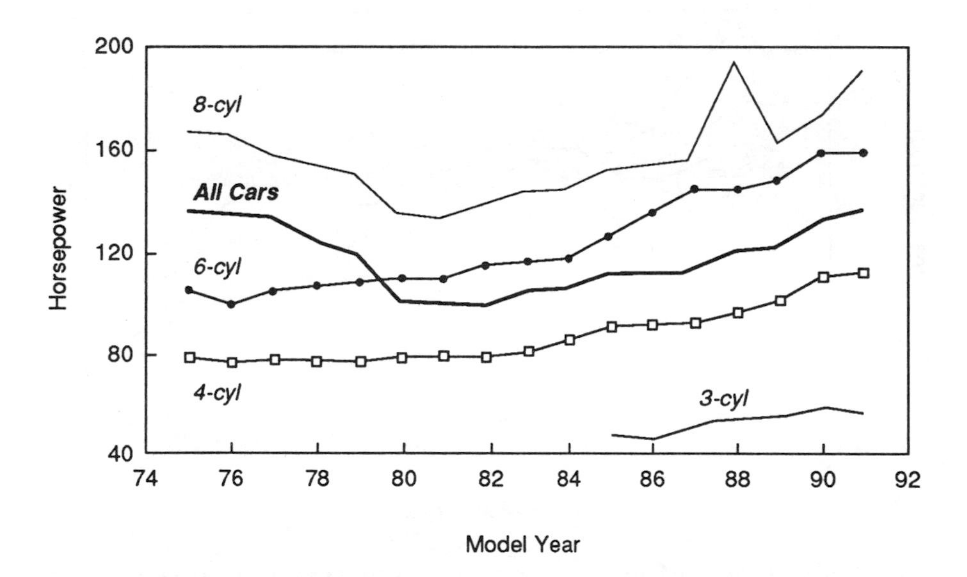

FIGURE 1-7 Average engine horsepower for passenger cars, 1975-1991.

SOURCE: Heavenrich et al. (1991).

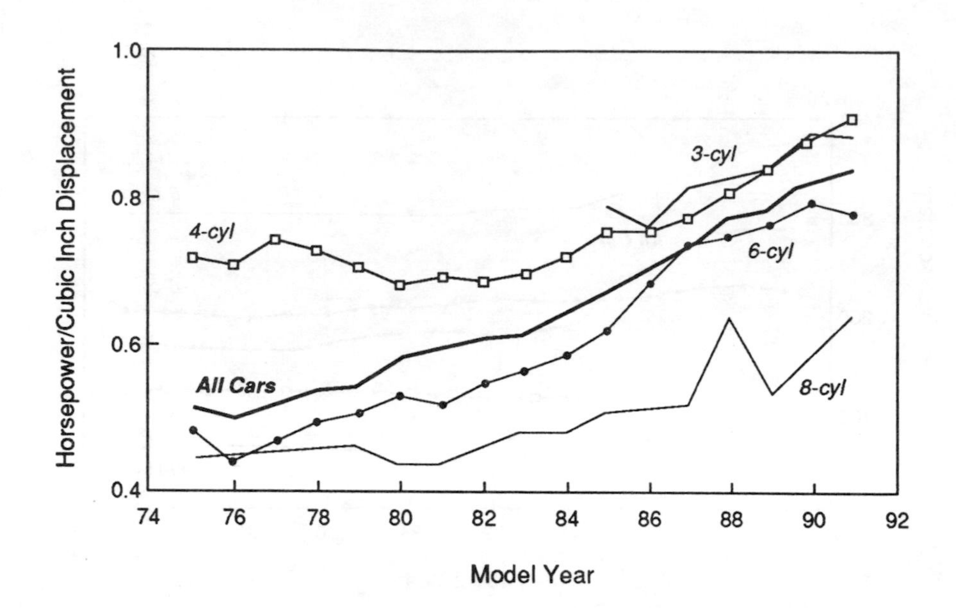

FIGURE 1-8 The ratio of horsepower to engine displacement for passenger cars, 1975-1990.

SOURCE: Heavenrich et al. (1991).

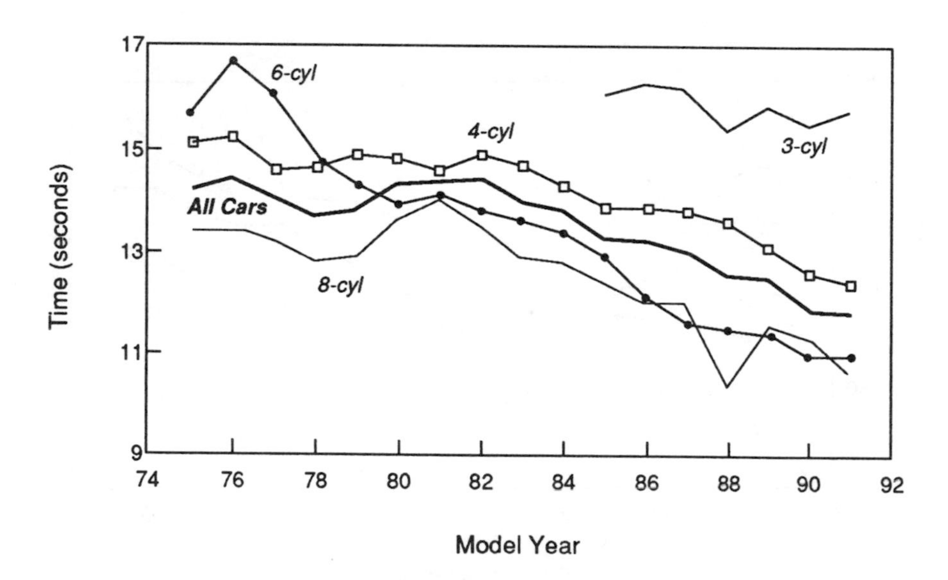

FIGURE 1-9 Performance of passenger cars as measured by time to accelerate from 0 to 60 miles per hour, 1975-1991.

SOURCE: Heavenrich et al. (1991).

Foreign automobiles have captured an increasing segment of the domestic market since 1975. The foreign share of the U.S. passenger-car market grew from 18.2 percent in 1975 to nearly 26 percent in 1990 (Motor Vehicle Manufacturers Association, 1991). The sales-weighted fuel economy of European imports has declined since 1982-1983, and that of Asian imports has declined since 1986 (Heavenrich et al., 1991). These trends are presumably a consequence of increasing sales of compact and midsize automobiles by foreign producers, as well as changes in performance, accessories, and other attributes. Thus, movement by consumers to foreign cars does not necessarily promise to improve fuel economy.

The fuel economy of light trucks has followed a path similar to that of automobiles (see Figure 1-1b). Light trucks, however, have lower CAFE requirements than automobiles and have not made the same dramatic gains in fuel economy. The fuel economy of light trucks improved by only 52 percent between 1975 and 1991 (13.7 to 20.8 mpg), compared with a 76 percent increase in that period by automobiles (15.8 to 27.8 mpg). As with cars, engine size decreased between 1975 and 1987 (Heavenrich et al., 1991). Average engine horsepower for light trucks reached a minimum in 1983, but has increased continually thereafter, which is in a direction contrary to improved fuel economy. Moreover, sales of light trucks have grown as a percentage of the light-duty vehicle market--from 19 percent in 1975 to 33 percent in 1991 (Figure 1-10). Because light trucks have lower fuel economy than automobiles, this trend has resulted in a decline in the combined fuel economy of the light-duty fleet to 25.0 mpg in 1991, the lowest level since 1985.

The extrapolation of the above trends into the future suggests that the fuel economy of the light-duty fleet is not likely to improve over the next several years. As noted above, the increasing demand for light trucks and greater performance adversely affects fuel economy, as does the increasing average weight of the new-car fleet. Moreover, the next several years will also see new requirements for additional safety equipment and improved emissions control, both of which will add weight and reduce fuel economy.[11]

In the absence of a supply interruption or some intervention to bring about a change in direction, significant reductions in fuel consumption are unlikely to be achieved in the future. Indeed, despite CAFE requirements, the aggregate annual consumption of gasoline by the fleet of light-duty motor vehicles has increased since the early 1980s (Figure 1-11).[12]

---

[11]Moreover, increasing road congestion and any tendency for consumers to keep their cars longer (and hence not to replace fuel-inefficient vehicles) will reduce the aggregate fleet fuel economy.

[12]Note that one barrel of petroleum typically yields about one-half barrel of gasoline and the equivalent of one-half barrel of other products. Refiners can adjust these proportions within broad limits, but only at a cost.

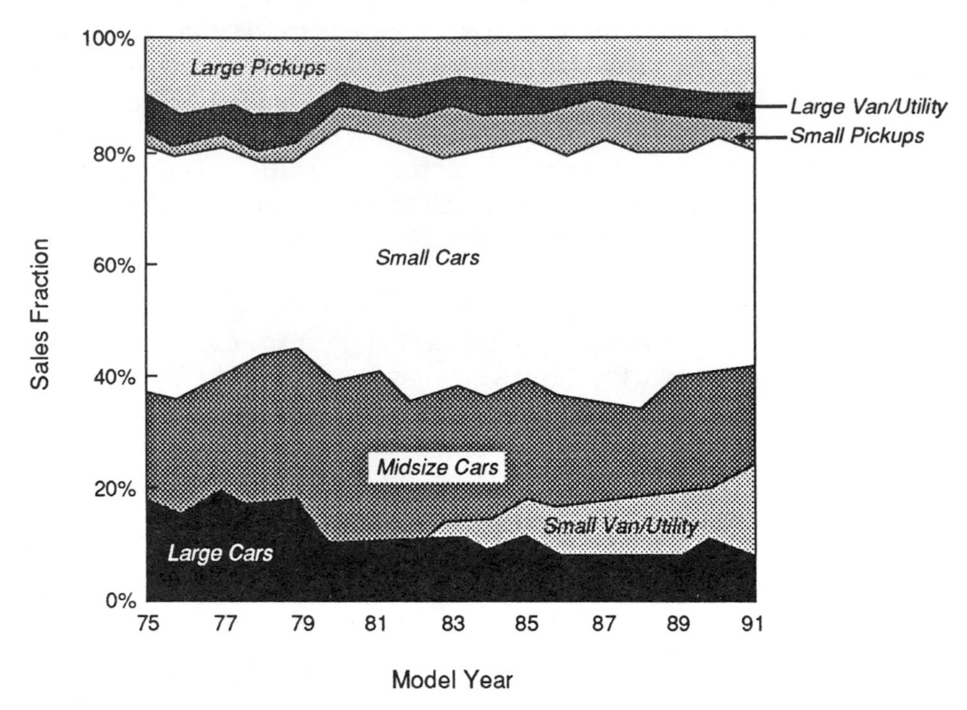

FIGURE 1-10 Car and truck sales by size class, 1975-1991.

SOURCE: Heavenrich et al. (1991).

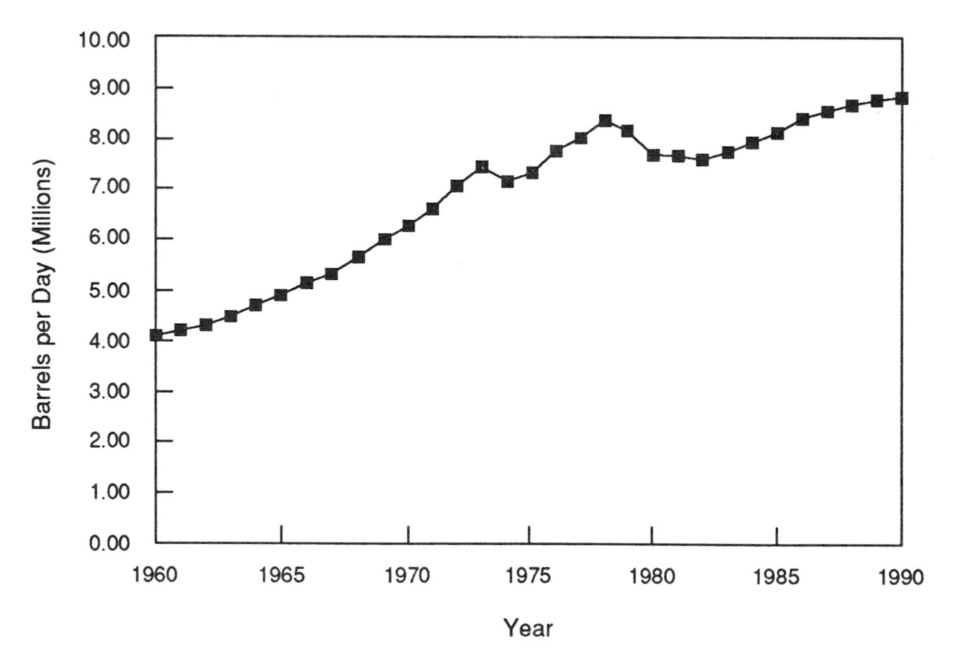

FIGURE 1-11 U.S. gasoline consumption, 1960-1990 (excluding small amount of diesel fuel used by some light trucks).

SOURCE: EIA (1991b).

THE COSTS AND BENEFITS OF REDUCED AUTOMOTIVE FUEL CONSUMPTION

A variety of costs and benefits are associated with lower fuel consumption by the light-duty fleet. These are briefly described below.

Potential Benefits

The motivation for limiting automotive fuel consumption stems from the conclusion that the reversal of the current upward trend in fuel consumption and petroleum imports presents benefits to society that are greater than the costs. The potential benefits include the following:

- **Reduced Consumer Expenditures.** Higher fuel economy will reduce vehicle operating costs and put downward pressure on gasoline prices, which could result in further savings to the consumer.

- **Conservation of Resources.** Petroleum is a depletable resource. The reduction of consumption would preserve petroleum for use by future generations.

- **Enhanced National Security.** U.S. concern with potential disruptions in its oil supplies from the major Middle East oil-producing regions has led to military and political involvement in the region. Any limitation of U.S. dependence on imported oil would presumably reduce the need for such activities. Reducing dependence on petroleum imports would also lessen the chances that embargoes could be successfully used as a "weapon" in disputes with the United States.

- **Enhanced Economic Security.** Reductions in fuel consumption would reduce the rate of increase and level of U.S. petroleum imports. This would help to reduce the vulnerability of the economy to disruption due to events in the oil-producing regions of the world. Reduced demand for oil by the United States would also lower fuel costs for consumers throughout the world and thereby provide an economic benefit to society.[13]

- **Improved Balance of Payments.** Current U.S. purchases of oil from abroad cost about $60 billion per year and, based on current trends, will increase substantially over the coming decades (EIA, 1991a,c). Reduced fuel consumption would reduce the level of imports and exert downward pressure on world oil prices, both of which would help to redress the U.S. balance-of-payments deficit.

- **Improved Environmental Quality.** Reduced fuel consumption would reduce emissions of carbon dioxide, thereby reducing the contribution of motor vehicles to the

---

[13] Several analysts have argued that the national security and economic benefits from reduced imports should be reflected by pricing oil higher than its market value. Estimates of the appropriate premium range from near $0 to over $100 per barrel of imported oil (Broadman, 1986).

production of greenhouse gases (GHGs).[14] Moreover, by reducing the total amount of gasoline that is consumed, hydrocarbon emissions could be reduced throughout the fuel cycle.[15]

- **Enhanced Diffusion of Technology.** Since the United States is such a large automotive market, strategies to increase the use of fuel-efficient technologies in the United States would help to diffuse such technologies globally. The United States would thereby provide important options for a world that will have to reconcile competing demands for energy, economic growth, and improved environmental quality.

- **Increased Economic Efficiency.** If fuel economy improvements can be introduced on a cost-effective basis--that is, if the costs of the technologies to improve fuel economy can be fully captured by reduced operating costs or other benefits--the overall economic efficiency and competitiveness of the U.S. economy would be enhanced.

Potential Costs

The achievement of the potential benefits of reduced fuel consumption may impose costs, depending on the strategy by which reduced consumption is achieved. The potential costs include the following:

- **Costs to the Consumer.** If the incremental costs of technologies to achieve increased fuel efficiency impose costs to the consumer that cannot be justified by the fuel savings, there is a net economic loss to the consumer. The new vehicle purchaser would have to pay more for a new vehicle, thus decreasing his or her purchasing power for other goods and services, or would have to buy a vehicle with fewer attributes than he or she would otherwise demand. Alternatively, the consumer could choose to continue operating an existing vehicle or to buy a used car, with resulting losses in consumer satisfaction.

- **Impacts on U.S. Automotive Manufacturers.** Requirements to increase fuel economy might require new expenditures by some U.S. automotive manufacturers at a time when they are confronting intense competition and are suffering serious financial losses. The increased costs would be passed along to consumers at least in part, which would lead to reduced sales and profits for the manufacturers. The extent and means by which increased fuel economy standards were implemented could also have important differential impacts among the manufacturers, favoring some and harming others. Moreover, automotive suppliers and related industries would also be affected.

---

[14]Gasoline-powered cars and light trucks account for about 57 percent of the GHGs generated by the U.S. transportation sector. This sector as a whole accounts for about 25 percent of total U.S. emissions of GHGs from anthropogenic sources (National Research Council, 1990).

[15]The fuel cycle includes the refinery-transportation-sales-vehicle system. Although improved fuel economy may not radically affect hydrocarbon emissions from automobile tailpipes, it will reduce the aggregate demand for gasoline and hence reduce the emissions from other parts of the cycle (DeLuchi et al., 1991).

● **Impacts on Employment.** The impacts on manufacturers would, of course, affect employment by the manufacturers and their suppliers. Auto workers already are confronting a significant number of plant closings.

● **Safety Impacts.** Fuel efficiency might be achieved by reducing vehicle size and/or weight. Moreover, increased manufacturing costs might be reflected in increased vehicle prices, which could result in some shifting of sales toward smaller vehicles. The impact of both of these effects on safety is not easily determined, however, because it depends on a complex set of factors.[16]

● **Conservation of Resources.** Resources spent on improving fuel economy are unavailable for other activities. Efforts to improve fuel economy standards beyond levels that are economically justifiable would divert resources from other sectors of the economy that might yield gains of greater societal value.

In sum, compelling benefits are associated with efforts to reduce petroleum consumption, but also significant costs. The United States should seek to adopt a national policy that achieves a balance of costs and benefits. Unfortunately, there is little national consensus on the magnitude and significance of the costs and benefits, and hence, there are disagreements as to how best to proceed. For example, some economists argue that the societal costs of the "externalities" associated with the use of gasoline (e.g., national security and environmental impacts) are reflected in the market price and that no additional efforts to reduce automotive fuel consumption are warranted.[17] Others argue that the externalities are substantial and that vigorous efforts are necessary and appropriate.[18]

The committee is not in a position to define the level of fuel economy that in fact achieves a balance of costs and benefits. That determination requires expertise and judgment that extend beyond its capability. Rather, the committee has sought to assess the costs that would be imposed at various levels of fuel economy. It is left to policymakers to assess whether the benefits of reduced fuel consumption are sufficient to justify the imposition of those costs.

---

[16]The relationship of size and weight to safety is discussed in Chapter 3.

[17]For example, Michael Boskin, chairman of the Council of Economic Advisers, presented this case to the committee during its workshop in Irvine, California, July 10, 1991. He stated, "It is generally known that economists favor internalizing externalities into the price structure rather than imposing a regulatory structure. You get near unanimity in the economics profession on that. With respect to the price of gasoline, the issue is really what the difference is between social cost and private cost. We already have a substantial amount of taxation at the Federal and State levels and there will be phased in increases in Federal gasoline taxes. . . . The Administration has no belief that externalities or extra social premiums that ought to be paid go beyond what's already on the books and scheduled to be implemented over the next year or so."

[18]Hubbard (1991) has presented a general case for full social-cost pricing of energy. Gordon (1991) discusses an analysis of this viewpoint for the transportation sector. There is a rich literature on estimating externalities, e.g., Bonneville Power Administration (1986); Chernick and Caverhill (1989); Freeman (1979); Hardin (1968); Kneese et al. (1970); Krutilla and Fisher (1975); Maler (1974).

THE COMPLEXITY OF THE PROBLEM

The assessment of the automotive fuel economy that is practically achievable involves a complex set of considerations. These considerations include the following:

- **Technology.** The identification of the technologies that could improve automotive fuel economy is one of the primary objectives of this study. Technology forecasting over a period of 15 years is, at best, an uncertain endeavor. The uncertainties include estimating the improvement in efficiency that a given technology will achieve, its cost, its interactive effects with other technologies, its societal impacts (e.g., emissions and safety), and its possible rate of penetration into the market. The longer the time horizon of the projection, the greater the uncertainties become. The evaluation is further complicated by the fact that some of the relevant information is proprietary.

- **Safety.** Some fuel economy improvements can be achieved without downweighting or downsizing. As greater fuel economy is sought, however, weight reduction or size reduction may become necessary. Moreover, as the price of vehicles increases, consumers may choose smaller vehicles. These decisions will have safety implications, although the impacts are not easily determined. Radically different inferences have been drawn by different researchers analyzing substantially the same data.

- **Emissions.** Automotive fuel economy improvements will reduce the quantity of carbon dioxide emitted per vehicle mile traveled and, by reducing the demand for hydrocarbons, will also reduce the hydrocarbon emissions from the entire fuel cycle. The impact of fuel economy improvements on the emission of other pollutants is not so readily assessed. The cost of new vehicles with higher fuel economy may slow the retirement of older, more polluting vehicles. Increasingly stringent emissions standards, especially for oxides of nitrogen, may preclude highly fuel-efficient technologies, such as advanced diesel, lean-burn, and two-stroke engines, from entering the market.

- **Manufacturers.** The domestic automotive industry is mature, cyclical, and competitive. Future profits will likely trail profits in the 1980s because of competitive pressures and expected lower demand for new vehicles. There is a serious question whether existing domestic manufacturers (and, indirectly, their suppliers) can afford the costs of stringent fuel economy standards while making the necessary investments to meet safety, emissions, and marketplace requirements. It is likely that domestic manufacturers will face severe challenges to their large-vehicle market by foreign competitors, especially those from Japan. Since domestic product lines dominate this market segment, competition in the segment will add to the financial pressures on domestic manufacturers. If the level and rate of change of fuel economy standards can only be met with expensive technology that significantly increases the price of cars, automotive sales will be reduced.

● **Employment.** The pressures on the domestic manufacturers and their suppliers also have consequences for their workers and are likely to lead to future reductions in domestic employment throughout the industry. Because the automotive sector is important to U.S. employment in other sectors, reductions in sales will have ripple effects throughout the economy. Stringent fuel economy requirements would aggravate these problems.

● **Consumer Behavior.** At the current level of gasoline prices, consumers do not consider fuel economy an important priority in their decisions to purchase new vehicles. Indeed, as discussed above, consumers are displaying an increasing interest in improved performance and light trucks. Historically, the preference for larger cars increases with the age of the buyer. Thus, projected U.S. demographic changes that point to a growing proportion of older people in the U.S. population may result in a trend toward larger, but not necessarily higher performance, cars. The need for manufacturers to satisfy consumer demand imposes pressures that conflict with improved fuel economy.

● **Regulatory/Legal Structure.** The impacts of policies to increase fuel economy are also affected by the form and nature of any regulatory requirements. The standards imposed may well have important differential effects among manufacturers and between domestic manufacturers and their foreign competitors.

ORGANIZATION OF REPORT

In the chapters that follow the committee evaluates the various factors outlined above. Chapter 2 describes the technologies that can result in higher fuel economy, including technologies not currently in wide use. Chapter 3 focuses on occupant safety and its relationship to vehicle downsizing and downweighting. Chapter 4 describes the emissions standards that cars and light trucks must meet and the impact of such standards on improved fuel economy. Chapter 5 describes trends in the automotive industry and the industry's viability, and Chapter 6 describes relevant consumer trends. Chapter 7 presents projections of possible future levels of "technically achievable" fuel economy based on several projection methods. Chapter 8 presents the committee's estimates of "technically achievable" levels of fuel economy by vehicle size class for 2001 and 2006, and discusses the technology, safety, emissions, environmental, and economic considerations that policymakers must consider in arriving at practically achievable levels of fuel economy. Finally, in Chapter 9, the committee comments on the current CAFE regulatory system and suggests improvements and possible alternative approaches.

REFERENCES

Bonneville Power Administration. 1986. *BPA Evaluation of Generic Environmental Costs and Benefits Studies*. Portland, Oreg.

Broadman, H.J. 1986. The social cost of imported oil. *Energy Policy* June:242-252.

Chernick, P., and E. Caverhill. 1989. *The Valuation of Externalities for Energy Production, Delivery and Use*. Boston: PLC, Inc.

DeLuchi, M., Q. Wang, and D.L. Greene. 1991. Motor vehicle fuel economy, the forgotten hydrocarbon control strategy. Oak Ridge National Laboratory, Tenn. In draft, June 18.

Energy Information Administration (EIA). 1990. *Annual Energy Outlook with Projections to 2010*. DOE/EIA-0383(90). Washington, D.C.: U.S. Department of Energy.

Energy Information Administration (EIA). 1991a. *Annual Energy Outlook*. DOE/EIA-0383(91). Washington, D.C.: U.S. Department of Energy.

Energy Information Administration (EIA). 1991b. *Annual Energy Review 1990.* DOE/EIA-0381(90). Washington, D.C.: U.S. Department of Energy.

Energy Information Administration (EIA). 1991c. *Monthly Energy Review*. DOE/EIA-0035(91/09). Washington, D.C.: U.S. Department of Energy.

Freeman, A. 1979. *The Benefits of Environmental Improvement*. Baltimore, Md.: Johns Hopkins University Press.

Gordon, D. 1991. *Steering a New Course: Transportation, Energy and the Environment*. Cambridge, Mass.: Union of Concerned Scientists.

Hardin, G. 1968. Tragedy of the commons. *Science* 162:1243-1248.

Heavenrich, R.M., J.D. Murrell, and K.H. Hellman. 1991. *Light-duty Automotive Technology and Fuel Economy Trends Through 1991*. Control Technology and Applications Branch, EPA/AA/CTAB/91-02. Ann Arbor, Mich.: U.S. Environmental Protection Agency.

Hubbard, H. 1991. The real cost of energy. *Scientific American* 264(4):36-42.

Kneese, A., R. Ayres, and R. d'Arge. 1970. *Economics and the Environment*. Resources for the Future. Baltimore, Md.: Johns Hopkins University Press.

Krutilla, J., and A. Fisher. 1975. *The Economics of Natural Environments*. Resources for the Future. Baltimore, Md.: Johns Hopkins University Press.

Maler, K.G. 1974. *Environmental Economics--A Theoretical Inquiry*. Baltimore, Md.: Johns Hopkins University Press.

Motor Vehicle Manufacturers Association. 1991. *Facts & Figures '91*. Washington, D.C.

National Research Council. 1990. *Confronting Climate Change: Strategies for Energy Research and Development*. Washington, D.C.: National Academy Press.

Oak Ridge National Laboratory (ORNL). 1991. *Transportation Energy Data Book*. llth ed. ORNL-6649. Oak Ridge, Tenn.

U.S. Department of Energy. 1991. *National Energy Strategy*. First Edition, 1991/1992. Washington, D.C.

U.S. Department of Transportation. 1991. *Briefing Book on the United States Motor Vehicle Industry and Market, Version 1*. Cambridge, Mass.: John A. Volpe National Transportation Systems Center.

2

FUEL USE IN AUTOMOBILES AND LIGHT TRUCKS

The fuel economy of an automobile or light truck is determined by its technology and design and by vehicle use, driver behavior and the conditions under which it is used--for example, speed, road design, weather, and traffic.[1] This chapter examines how the design and technical characteristics of an automobile or light truck affect its fuel economy, with a focus on technologies that might be employed to achieve fuel economy gains in the future.[2]

High fuel economy is only one of many desirable vehicle attributes. Consumers also value acceleration and handling, safety, comfort, reliability, passenger- and load-carrying capacity, size, styling, and low noise and vibration, not to mention low purchase and ownership costs. Society at large requires vehicles to have certain additional attributes, such as low exhaust emissions. All of these attributes influence vehicle design and technology, and most of them affect fuel economy. As a consequence, the fuel economy of a vehicle results from the trade-offs, guided in part by costs and benefits, that must be made among a variety of vehicle characteristics.[3]

---

[1]To control for the effects of the variables that are external to the vehicle, it is customary to evaluate the fuel economy of a vehicle using the Environmental Protection Agency's (EPA's)Federal Test Procedure (FTP), which involves laboratory measurement of fuel economy over specific simulated urban and highway driving cycles.

[2]It was beyond the scope of this study to examine the determinants and effects on fuel consumption of vehicle use (typically measured by vehicle miles traveled, or VMT), driver behavior, and driving conditions. All contribute importantly to the total national consumption of gasoline, diesel, and other fuels for automobiles and light trucks.

[3]It is possible to produce vehicles with extraordinarily high fuel economy, but only at the sacrifice of nearly all other important vehicle characteristics. For example, an experimental passenger-carrying vehicle, optimized for fuel economy alone, has achieved a world record 6,409 miles per gallon (mpg) (Associated Press, 1988). However, it can carry only a child or small adult in the prone position, and can achieve a speed of only 15 miles per hour (mph) on level ground. It has very limited hill-climbing ability, would fail federal safety tests, and might fail emissions tests as well. An automobile with maximized fuel economy would simply be unusable and unacceptable to consumers.

Changes over the years in vehicle design and in automotive technologies have made it possible to achieve substantial increases in fuel economy while maintaining acceptable levels of other attributes, including price. This chapter illustrates some of the fundamental principles underlying the fuel economy characteristics of automotive technologies. It also identifies some of the improved technologies that are available now or that are in the R&D stage.

The details of how some of the technologies discussed in this chapter might contribute to fuel economy improvement are presented in Appendix B, and their implications for size-class and fleet fuel economy are discussed in Chapters 7 and 8. Certain emerging technologies--especially alternative engine designs--are described later in this chapter and discussed further in Appendix C.

THE VEHICLE AS A SYSTEM

In any vehicle that is propelled by burning fuel in an engine, the chemical energy in the fuel is released by combustion to fulfill a number of functions. Part of the fuel's energy is lost to the vehicle's surroundings as low-grade (low-temperature) heat in the exhaust gas and the air that cools the engine and radiator. Part of the chemical energy is transformed into mechanical energy, or *work*. A portion of the work propels the vehicle by overcoming (a) inertia (weight) when accelerating or gravity when hill climbing, (b) the resistance of the air to the motion of the vehicle, and (c) the rolling resistance of the tires on the road. Another portion operates the vehicle's essential and optional accessories. Yet another portion operates the engine itself and overcomes the frictional energy losses that occur in every part of the vehicle system.

The proportions of the chemical energy that are used to fulfill each of the several functions for a typical vehicle of modern design are not fixed. Rather, they depend on the details of the vehicle's design and the nature and efficiencies of the vehicle's technologies. Improvements in fuel economy are usually accompanied by changes in the proportion of the fuel's energy that goes to satisfy each function. For example, the energy required to propel the vehicle can be decreased by reducing inertia, the drag imposed by the air on the moving vehicle, or the rolling resistance of the tires. The energy required to operate accessories can be reduced by changing their design, increasing their efficiency, or, in the case of optional accessories, eliminating them. The energy required to operate the engine itself can be reduced by adopting different engine technology or making improvements in the detailed engineering design of the engine. The energy lost to friction in the rest of the vehicle system can be reduced by making any of a large number of design changes and technical improvements in, for example, the transmission and other parts of the drivetrain or by improving lubrication of their elements.

It is also not possible to attach precise quantitative values to each of the energy flows because the distribution varies greatly depending on operating conditions. A vehicle that is optimized for fuel economy at highway speeds might have unacceptable performance in the stop-start usage of urban travel. In recognition of this fact, the FTP covers a wide range of operating conditions, including cold start, idle, acceleration, deceleration, and cruise at low and high speeds. Consequently, the measured fuel economy (and emissions performance) is a composite of energy use under a variety of operating conditions.

Automotive designers and engineers must optimize the vehicle and its power train to meet the often-conflicting demands of customer satisfaction, fuel economy requirements, emissions standards, and cost in the variety of operating conditions under which the vehicle will be used. Complete consideration of the effects on fuel economy of changes in design or technology requires examination of the vehicle as a total *system*. For example, reducing the size of the engine may reduce the vehicle's weight and thus reduce the fuel needed to accelerate the vehicle and propel it up hills. But, unless engine performance is improved or the drivetrain is altered, the smaller engine may accelerate the vehicle less rapidly. If those changes are introduced, however, the "feel" of the vehicle will change, with associated impacts on consumer acceptance. Modifications of technology change the system and hence have complex effects that are difficult to capture and analyze.

It is usually possible, however, to estimate the impacts of specific technologies in terms of a percentage savings in fuel use for a typical vehicle without a full examination of all the system-level effects. This approach is used in most of this report. The analysis in Chapter 7 of the fuel economy potential of future vehicles takes into account some of the most important systems implications for fuel economy of certain alternative technologies. Nonetheless, changes in technology have both obvious and subtle effects that could prove significant, and the committee's analysis cannot capture all of these impacts.

ENGINE TECHNOLOGY AND FUEL USE

Engine design and technology are key determinants of the fuel economy of an automobile or light truck. The engine's efficiency in converting fuel energy to useful work affects fuel economy directly, and its design influences most of the other characteristics of the vehicle, especially its weight and size, which further influence fuel economy.

Standard Engine Technology

The standard engine in today's vehicles is a spark-ignition, four-stroke, internal combustion engine of four, five, six, or eight cylinders.[4] The basic concepts underlying this engine have been understood and applied for over a century. Yet, improving the efficiency, operability, and emissions characteristics of this engine is the subject of continued fruitful study and evolutionary innovation even today.

Many variations on the standard engine design have been used in vehicle applications. Some important variations, which may be used in combination, include the multivalve engine (which employs two valves or more per cylinder for intake of the air/fuel mixture and for exhaust of the combustion products), the variable-valve-timing engine (in which the timing of valve operation can be varied with respect to the crankshaft position), the two-stroke engine (which combines the functions of the four strokes into two), and the lean-burn engine (which burns the fuel in the presence of excess air and/or exhaust gases and which may also include alternative configurations for, and precise control of, the pattern of combustion of the fuel). These variations, and the details of design within each, offer the possibility of different and improved combinations of fuel economy, performance, and cost.

The compression-ignition, or diesel, engine differs in several ways from the spark-ignition engine, although it is based on the same two- or four-stroke cycle. In the diesel, air alone is compressed until late in the compression stroke, when fuel is injected into the cylinder at very high pressure. Because the fuel is injected late in the cycle, a diesel engine may use a much higher compression ratio. This, combined with its ability to burn a "lean" air/fuel mixture (see below), makes the diesel more efficient than a spark-ignition engine. On the other hand, the diesel engine is typically heavier than an equivalent spark engine, and it produces a different set of exhaust emissions. (See Appendix C for further discussion of diesel technologies.)

In a spark-ignition, four-stroke engine of standard design, a portion of the energy released in combustion does work as the hot, high-pressure combustion gases expand

---

[4]In such an engine, a mixture of fuel and air is delivered by a carburetor or fuel-injection system to the cylinders, where it is compressed by pistons moving up in the cylinders (the *compression* stroke) and is ignited by an electrical spark. As the fuel burns, it produces a mixture of hot, high-pressure gases whose controlled expansion in the cylinders causes the pistons to move down (the *expansion* stroke). Through a mechanical linkage, this downward motion of the piston in the cylinder is translated into a rotating motion of the crankshaft. The piston then moves upward again (the *exhaust* stroke) and, with the simultaneous opening of the exhaust valve on the engine, the burned fuel products (exhaust gases) are pushed out of the cylinder and into the exhaust manifold. Finally, the piston again moves down (the *intake* stroke), with the inlet valves open this time to admit fresh fuel/air mixture. When the piston reverses motion, it begins again to retrace the sequence starting with the first stroke. In this way, a portion of the fuel energy released during engine operation is transformed into mechanical energy of the rotating crankshaft, which is then available to propel the car, operate accessories, and overcome frictional losses in the drivetrain.

All the pistons are attached to a common crankshaft in such a way that the expansion stroke of each cylinder occurs at different times. Thus, the torque (twisting force) transmitted to the crankshaft from the pistons through steel shafts, or connecting rods, is more nearly uniform in time than it would be if all the pistons were in the expansion stroke at the same time. The greater the number of cylinders, the more uniform is the torque on the crankshaft and the smoother is engine operation.

and force the piston down during the expansion stroke.[5] However, not all of this work is available for moving the vehicle, in part, because a portion of it must be used to compress the air/fuel mixture during the compression stroke. The difference between the work made available during the expansion stroke and that used during the compression stroke is conventionally defined as the *indicated work*. A portion of the indicated work--the *pumping loss*--is required to pump the air and fuel into the cylinders during the intake stroke and to pump the exhaust gas out during the exhaust stroke. Yet another portion of the indicated work is transformed into heat by friction among the moving parts of the engine. The remaining work, called the *brake work*, is available at the engine crankshaft. This brake work, less friction losses in the drivetrain and accessory losses, is transferred by the drivetrain to the wheels of the vehicle. Figure 2-1 illustrates the pathways for the use of the indicated work during vehicle operation on a level road.

Each of the energy pathways offers the potential for fuel economy improvement. For example, fuel economy might be increased by redesigning the vehicle exterior to reduce the aerodynamic drag forces that consume a substantial portion of the fuel energy at high speeds. Or, fuel economy might be increased by reducing friction in the engine, drivetrain, accessories, or other parts of the vehicle through use of different technologies or better lubricants.

The indicated work, pumping loss, and friction loss vary with the load (see Figure 2-2) and, to a lesser extent, the engine speed. At decreased load, a greater fraction of the indicated work goes to overcoming pumping loss and friction. Consequently, reducing pumping losses and engine friction will significantly improve fuel economy for engines that operate at light load much of the time. Figure 2-2 also illustrates why driving in a higher gear at a given vehicle speed generally improves fuel economy--in a higher gear, the engine runs at lower revolutions per minute (rpm), where friction is lower, and at a higher load, where pumping and friction losses are proportionately smaller.

---

[5]*Indicated thermal efficiency* is usually defined as the proportion of the energy of the fuel that is converted to work and delivered to the top of the piston during the compression and expansion strokes. It is determined primarily by the rate of burning, heat losses, compression ratio, and air/fuel ratio.

The rate of burning is affected by in-cylinder turbulence, combustion chamber design, spark plug location, and air/fuel ratio. A high rate of burning is desirable, but, if excessive, it can cause undesirable engine noise, vibration, and harshness. Cooling of the air/fuel mixture late in the compression stroke minimizes the tendency of the engine to "knock" (premature burning and detonation in the cylinder), but cooling of the exhaust gases after combustion decreases the work obtained during expansion.

The compression ratio is limited in the spark-ignition engine by knock, which is affected by fuel octane rating, combustion chamber design, spark timing, combustion chamber cooling, and deposits on the combustion chamber wall. Within limits, knocking can be controlled by using sensors to detect it and by delaying, or retarding, the spark--an action that decreases fuel economy.

The air/fuel ratio, with a few exceptions, is fixed by emissions considerations because the catalytic system used in modern vehicles requires a stoichiometric air/fuel mixture (i.e., a mixture that contains precisely the correct amount of oxygen needed for complete combustion of the fuel). There is a fuel economy advantage, but an emissions problem, when excess air is used, as discussed in Appendix C. For homogeneous mixtures of air and fuel, the rate of burning and combustion stability decrease on either side of an approximately stoichiometric mixture, so there is a practical lean-combustion limit for homogeneous mixtures. This limit does not apply to lean, stratified-charge engines, such as the diesel. (See Chapter 4 and Appendix C.)

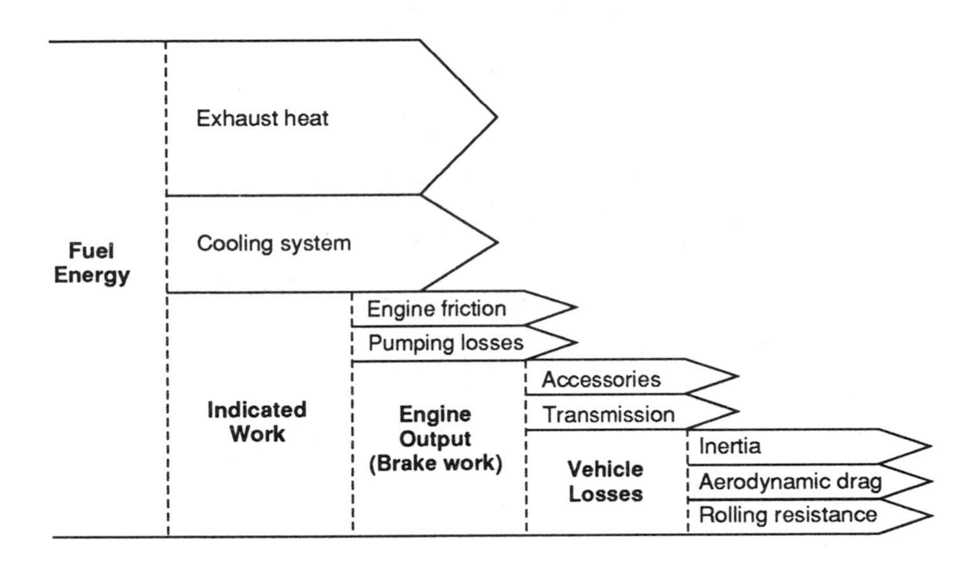

FIGURE 2-1 Where the energy in the fuel goes (proportions vary with vehicle design and operating conditions).

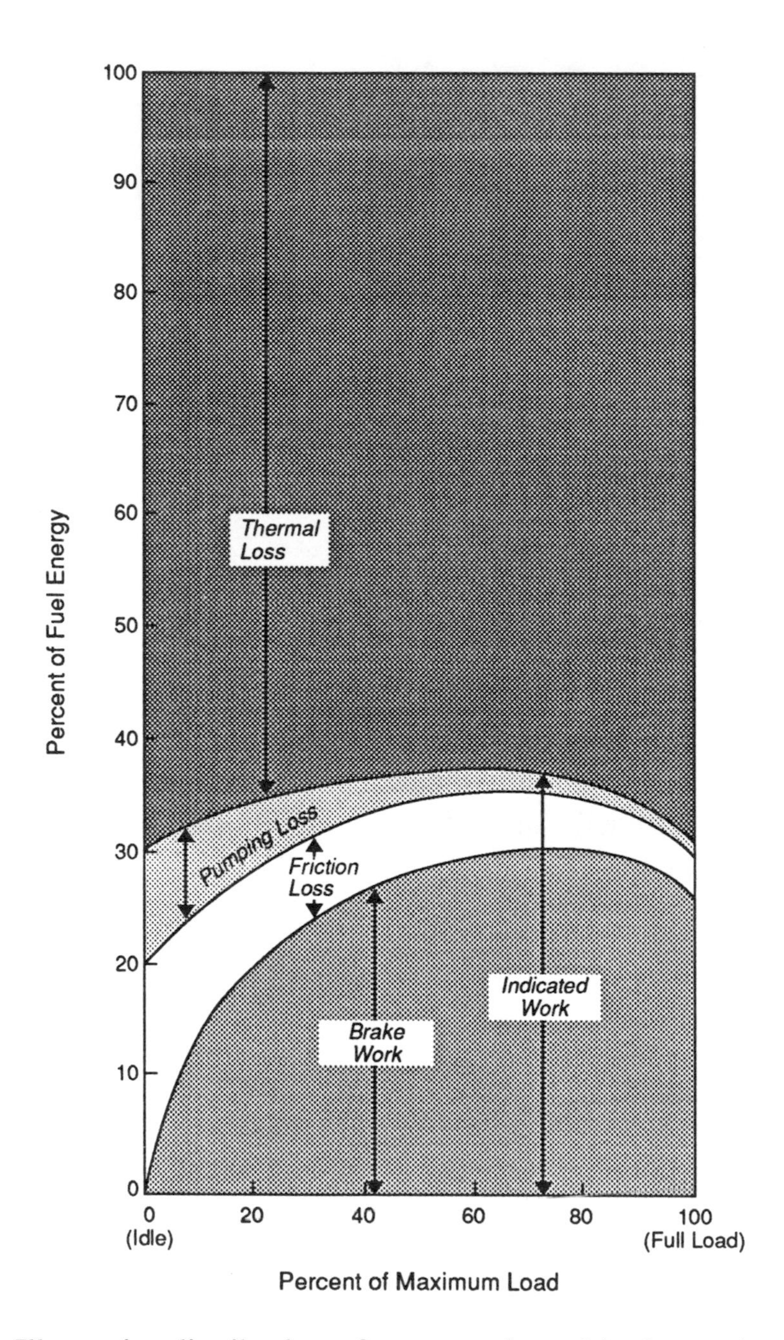

FIGURE 2-2 Illustrative distribution of energy released in the engine as a function of load.

SOURCE: Adapted from Amann (1991).

Table 2-1 shows how the brake work delivered to the drive wheels is distributed among the three major categories of force (aerodynamic drag, rolling resistance, and inertia) against which the engine must work during three different "cycles" of driving, all on a level road. One cycle is driving at a steady speed of 55 to 65 mph. The other two cycles are the urban and highway parts of the EPA's FTP. The urban cycle includes numerous periods of acceleration and deceleration, accompanied by periods of moderate, steady speeds. The highway cycle includes fewer acceleration and deceleration periods and more steady driving at speeds higher than in the urban cycle, although at an average speed less than the 55 to 65 mph of the steady-state cycle.

As shown in Table 2-1, the proportion of the energy used to overcome aerodynamic drag increases substantially at higher speeds. (Drag increases as the square of vehicle speed.) The proportion of the energy used to overcome inertia is greatest during the urban cycle, with its frequent stops and accelerations.

Performance/Fuel Economy Trade-Offs

One of the more significant trade-offs in automotive design arises from the tendency for high performance to require more fuel. For example, engine design can affect both fuel economy and the capability of a vehicle to accelerate or climb hills quickly. A vehicle's acceleration performance is directly related to the torque exerted on the driving wheels, which is proportional to the engine torque conveyed through the transmission and drivetrain. Thus, deliverable torque, which depends on both engine speed and load, is a significant parameter of engine performance.

TABLE 2-1 Illustrative Distribution of the Energy Delivered to the Drive Wheels

	Operating Conditions		
Force to be Overcome	Steady-State Cycle (55-65 mph)	EPA Highway Driving Cycle	EPA Urban Driving Cycle
Aerodynamic drag	75-80%	50-55%	28-25%
Rolling resistance	25-20	35-40	28-25
Inertia	0	15-5	44-50
Total	100%	100%	100%

SOURCE: MacCready (1989).

For a given engine design and air/fuel ratio, the torque produced by an engine increases as greater amounts of air are introduced into its cylinders during the intake stroke. The flow rate of air can be increased by increasing the size of the engine or by introducing air at higher pressures so as to increase its density. (Engine size is measured by its *displacement*, that is, the volume swept by the pistons during one stroke.) Increased engine size increases engine friction and reduces fuel economy. Efforts to increase air density may have similar effects.[6]

Engines with different designs but the same displacement may have very different curves of engine torque versus engine rpm, as illustrated in Figure 2-3, which was prepared by Honda Motor Company to explain the torque characteristics of its new VTEC engine (see Appendix C). The VTEC engine has a mechanism that permits it to change from 2-valve to 4-valve operation at the shift point, as shown in Figure 2-3. The valve timing also differs between 2- and 4-valve operation.

As shown in Figure 2-3, for a 2-valve engine, the combination of valve timing and increased flow restriction creates a torque peak at about 2,300 to 2,400 rpm. However, a 4-valve engine with different valve timing could have a torque curve that peaks near 4,800 rpm. Honda has designed its valve train for 2-valve operation at low engine speed and for 4-valve operation at high engine speed. This produces the upper torque curve for the VTEC, which is relatively flat over a wide range of engine speed.

Thus, used in the same vehicle, the VTEC engine, which delivers higher horsepower as well as higher torque at the same engine speed, delivers higher acceleration performance (and probably equal or higher fuel economy) than the standard Honda engine with the same displacement. Alternatively, the VTEC engine

---

[6]Increasing the ambient air density increases the mass of air in the cylinder and consequently increases torque and power output. One way to increase ambient air density is to connect an air compressor to the intake manifold. The power required to drive the air compressor can come from the engine crankshaft (using a supercharger) or from a gas turbine driven by the engine exhaust (a turbocharger). The supercharger is not activated until wide-open throttle is reached, and when it is activated, it momentarily reduces the power available to move the vehicle. (Because of the benefits of increased air density, however, the net effect is a power increase.) The response of a supercharger to sudden power demands is instantaneous, but there is a small delay in the response of a turbocharger because of the delay in the exhaust-gas flow rate and the need to overcome the inertia of the turbocharger. A variable-speed drive between the crankshaft and the supercharger is sometimes used because of the mismatch between the flow characteristics of the reciprocating engine and the rotating compressor usually used. A variable-displacement compressor has also been proposed for the same purpose.

There is considerable disagreement about the effects of these arrangements on fuel economy. When ambient air density is increased, a smaller displacement engine can be used to produce the required power. The smaller engine should have less friction and may be lighter, both of which would improve fuel economy. However, the higher effective compression ratio that accompanies use of turbo/supercharging results in an increased tendency for the engine to knock and may require the use of knock sensors to retard the spark. Retarding the spark is detrimental to fuel economy. Also, some engineers believe that the higher cylinder pressures encountered when ambient air pressure is increased will require, for equal durability, a stronger engine structure, which in turn, will increase engine weight and engine friction and decrease fuel economy. As a result of this controversy, the committee did not take turbo- or supercharging into account in its analyses, even though both are in limited use today to enhance vehicle performance.

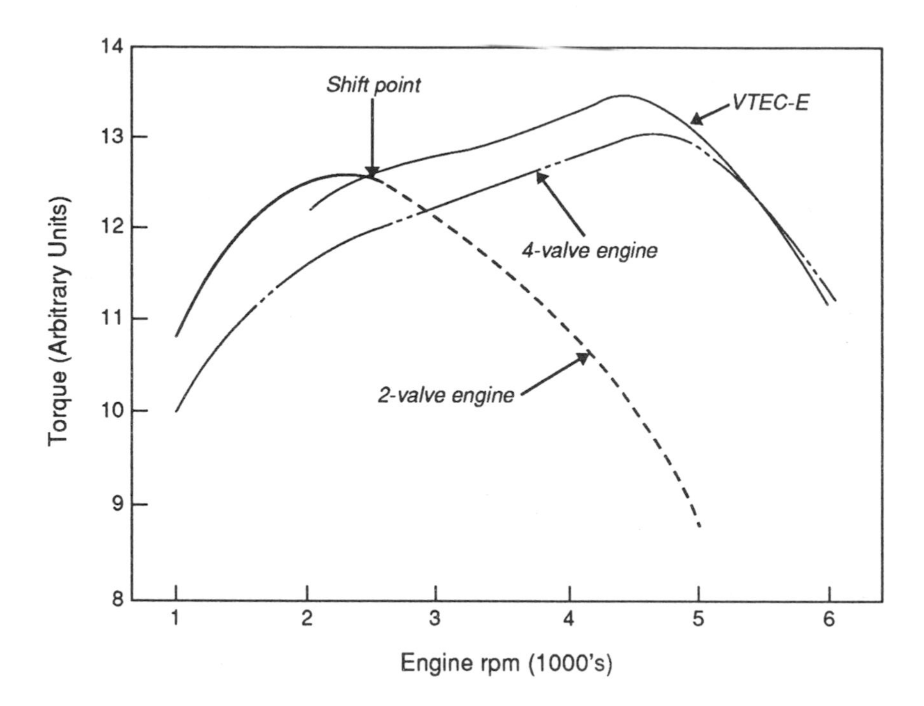

FIGURE 2-3 Torque curves for various engines.

SOURCE: Honda Motor Company, Ltd. (1991).

could be accompanied by changes in the drivetrain to give increased fuel economy rather than higher acceleration performance. In short, a range of fuel economy performance can be obtained not only from different engines with the same displacement, but also from the same basic engine.

The 4-valve engine also provides an excellent example of how a technology can be used for different purposes. Most, if not all, implementations of the 4-valve engine to date have been to improve performance rather than fuel economy. Numerous design details of engines and drivetrains could be changed to maximize vehicle fuel economy rather than performance.

Technologies to improve fuel economy can also affect performance attributes other than acceleration. Downsizing an engine from six to four cylinders may improve fuel economy, but at the cost of increased engine-induced vehicle vibration. Although a 5-speed transmission weighs more than a 4- or 3-speed transmission, by allowing more efficient engine operation it may yield both better performance and higher fuel economy. Substituting a continuously variable transmission for an automatic one also can save fuel, but with modifications to the "feel" of a car in ways that consumers may not like. However, not all technologies to improve fuel economy necessarily require compromises of other performance attributes. For example, multipoint fuel injection, which enables more precise fuel metering and control, not only improves fuel economy, but also can yield lower emissions and improved responsiveness. As a result, a substantial proportion of all new passenger cars use multipoint fuel injection, despite the fact that this technology may not be cost-effective as a fuel economy technology alone (SRI, 1991).

FUEL ECONOMY TECHNOLOGIES FOR THE NEXT DECADE

Table 2-2 lists the two categories of technologies--*proven* and *emerging*--that the committee determined to be important to the fuel economy of future automobiles and light trucks over the period to 2006. The selection was based on published studies and presentations to the committee (Berger et al., 1990; DeCicco, 1991; Energy and Environmental Analysis, 1991a,b; Ledbetter and Ross, 1990; Office of Technology Assessment, 1991; SRI, 1991).

Proven Technologies

Many concepts, technologies, and designs have the potential to improve the fuel economy of motor vehicles. *Proven* technologies are ones that can reasonably be expected to be available for incorporation in the future vehicle fleet by automotive manufacturers. To be considered "proven" by the committee, a technology had to be currently available in at least one mass-produced light-duty vehicle. An emphasis on proven technology is warranted because of the need for assurance that a fuel economy gain can in fact be achieved by manufacturers, who must be concerned with safety, emissions, reliability, cost, manufacturability, and consumer acceptance, as well as fuel use.[7]

The proven technologies vary widely in their potential contribution to increased automotive fuel economy, their costs, and the degree to which they are used in mass-produced vehicles today. There is a considerable divergence of views in the literature about these aspects of a number of the proven technologies. The committee examined each of the technologies at length, and the details of that inquiry are presented in Appendixes B and E. Appendix B includes (1) a discussion of the technology and the aspects of vehicle energy use it affects, (2) estimates of the improvement in fuel economy that may be achievable, compared with a baseline technology, and (3) estimates of the cost of using the technology. Among other data, Appendix E includes estimates of the market share of each of the proven technologies in the current fleet and the committee's judgments about the potential future market share of each.

---

[7]Many other technologies have been demonstrated to the public in prototype and concept vehicles. The constraints applied during the design of concept vehicles are different from the constraints applied during the design of production vehicles. Consequently, fuel economy values obtained from concept vehicles are not directly useful in estimating the potential fuel economy of production vehicles (see Appendix C).

TABLE 2-2 Fuel Economy Technologies for Automobiles and Light Trucks

Proven technologies[a]

Engine technologies
- Roller cam followers
- General friction reduction
- Deceleration fuel restriction
- Compression ratio increase
- Throttle-body fuel injection
- Multipoint fuel injection
- Overhead camshaft
- Four valves/cylinder
- Variable valve timing
- Four-cylinder engine
- Six-cylinder engine
- Advanced lubricants
- Turbo/supercharging[b]
- Diesel[c]

Transmission technologies
- Torque converter lockup
- Electronic control
- Four-speed automatic
- Five-speed automatic
- Continuously variable transmission
- Five-speed manual

Proven technologies (continued)

Accessories
- Accessories
- Electric power steering

Rolling resistance
- Advanced tires

Inertia
- Weight reduction
- Front-wheel drive

Aerodynamic drag
- Aerodynamics

Emerging technologies

- Lean-burn engine
- Two-stroke engine
- Active noise control
- Lean NO_x catalyst

NOTE: Descriptions of the technologies and their fuel-use characteristics, along with estimates of their costs, are provided in Appendix B.

[a]The categories of proven technologies correspond to the energy pathways in Figure 2-1.

[b]Currently used primarily to enhance performance. Because their ability to improve fuel economy is controversial, these technologies were not included in the analyses (see footnote 7).

[c]Light trucks only. Discussed in Appendix C as an emerging technology.

The proven technologies list includes technologies that will be familiar to most American consumers who are attentive to the new car and truck markets in the United States, including mass-market imported vehicles. Better engines, more efficient transmissions, body designs with improved aerodynamics, and lighter weight vehicles are all staples of the sales and marketing activities of major automobile and light-truck producers. On the other hand, most of these technologies have reached only a fraction of their potential application in vehicles sold in the United States, and as detailed in Chapter 7 and Appendix E, rather substantial increases in new-car and light-truck fuel economy by vehicle type and class could be achieved if they were to be employed to their maximum potential.

A substantial portion of the proven technologies relates to changes in the technology of the standard four-stroke, spark-ignition engine or to use of transmissions with larger numbers of gear ratios (including the continuously variable transmission, which conceptually has an unlimited number of ratios). The discussion of the standard engine earlier in this chapter provides the fundamentals of the technical background for understanding how each of these technologies works and why it is potentially important to fuel consumption. More details on the fuel economy aspects of these technologies are given in Appendix B.

It is difficult to compare the different published costs for each proven technology because the costs are usually reported as the marginal costs above some baseline technology, and the selection of the baseline for comparison can make a large difference in the reported cost. The same issue arises in examining the potential fuel economy contributions of a technology. This matter is examined in detail for the proven technologies in Appendix B, and all of the numbers reported there have been put on as consistent a basis as possible by the committee. The choice of a baseline for comparison is a somewhat arbitrary matter--for example, in choosing the type of transmission to use as the base for comparison with the other types. The choice can also reflect the more fundamental question of whether a technology is treated as an addition to, or replacement of, part of an existing vehicle already in production, or whether it is incorporated in an entirely new design.

Some proven technologies offer the promise of major increases in fuel economy for cars and light trucks, yet their costs are also quite high. Typical of these are such technologies as multipoint fuel injection, engines with four valves per cylinder, and vehicle weight reduction. Others appear to offer relatively limited fuel economy benefits, yet their costs are relatively low as well. In making decisions about which of the proven technologies to incorporate in a vehicle, manufacturers and their customers, if acting on a financially reasonable basis, will adopt proven technologies in the order of their cost-effectiveness--that is, in the order in which they provide the most fuel economy improvement for each dollar invested in them. Beyond higher fuel economy, some proven technologies offer other desirable attributes to producers and consumers, such as enhanced acceleration performance or better control over emissions. Multipoint fuel injection is such a technology--it has been adopted for many makes and models of vehicles, even though on fuel economy grounds alone it may not be cost-effective.

The precise ordering of new technologies according to their cost-effectiveness is sensitive, then, to the cost and potential for fuel economy improvement of each one. These factors differ not only among authors, but also among vehicle types and sizes, as explained in Appendixes B and E. Thus, it is not possible to make generalized statements about which technologies are the most cost-effective in the general case. Therefore, for each technology, the committee carried out a detailed examination of its cost, fuel economy improvement potential, and cost-effectiveness for cars and light trucks of different sizes and types, using data from different sources as the basis for the examinations. The implications of this analysis for fuel economy improvements within vehicle size classes are discussed in Chapter 7.

Emerging Technologies

The committee defined emerging technologies as those that hold promise of improving fuel economy, but whose future success is uncertain. Because a number of such technologies are being studied, and intense development is under way at a number of places, the driving forces toward successful application are strong. Thus, it seems reasonably certain that one or more emerging technologies (probably including some not foreseen in this report) will begin to make significant contributions to fuel economy by the year 2006.

Two widely discussed emerging technologies, lean-burn and two-stroke engines, exemplify the types of new engine technologies that may offer promise of enhanced fuel economy in the long term. In the committee's judgment, however, neither alternative is likely to achieve widespread practical vehicle application during the next decade, despite current commercial uses. Technical details of their operation, their fuel economy potential, and barriers to their adoption, including environmental limitations, are discussed in Appendix C.

Lean-burn engines are designed and operated so that excess air over and above that needed for complete combustion is introduced into the combustion chambers.[8] Because of its potential for increased fuel economy, the homogeneous lean-burn approach was investigated extensively in the 1960s and early 1970s as an alternative emissions control approach to the three-way catalyst, which requires use of a stoichiometric air/fuel mixture that leads to lower fuel economy.[9] However, the lean-burn engine was unable at that time to meet emissions and drivability requirements and its development was discontinued. Its current revival is due to its acknowledged fuel economy advantage and the availability of electronic fuel injection, which makes possible operation under lean-burn conditions in selected portions of the driving cycle. The difficulty in limiting NO_x emissions is a major problem with lean-burn engines, however. A lean NO_x catalyst, discussed in detail in Chapter 4, would markedly

---

[8]*Lean burn* is also sometimes used to describe an engine in which exhaust gases, rather than excess air, are used to dilute the air/fuel mixture. The diesel engine, discussed earlier and detailed in Appendix C, is a stratified-charge, lean-burn engine.

[9]A three-way catalyst allows the simultaneous control of unburned hydrocarbons (HC), oxides of nitrogen (NO_x), and carbon monoxide (CO).

improve the potential for use of lean-burn engines, but much fundamental research remains to be done in this area.

Two-stroke engines use an air pump other than the engine's pistons and cylinders to help accomplish the four tasks of compression, expansion, intake, and exhaust in only two strokes (one revolution of the crankshaft). In the two-stroke engine, exhaust and air (or air/fuel) intake are accomplished late in the expansion stroke and early in the compression stroke, respectively, by blowing the compressed air or mixture into the cylinder through the intake valve or port. Ideally, all of the products of combustion (and none of the incoming gases) would be blown out of the open exhaust valve or port, leaving primarily fresh gases in the cylinder. In practice, however, a significant portion of the compressed input mixture escapes through the exhaust valve, and a significant portion of the exhaust gases remains in the cylinder. If the incoming gases contain fresh fuel, the exhaust will contain unburned fuel, leading to high levels of unburned HC in the exhaust.

The potential benefits from automotive applications of two-stroke technology include reduced engine weight, size, and cost. Further, since each cylinder undergoes a power stroke on every revolution of the crankshaft rather than on every second, for the same number of cylinders a two-stroke engine generates somewhat less than twice the output power and operates more smoothly than a four-stroke engine. If the vehicle is optimized around a two-stroke engine, there is the potential for improved fuel economy.

Significant problems related to mechanical components and exhaust emissions must be overcome, however, before the two-stroke engine can be a serious competitor to the four-stroke engine. Although NO_x emissions are reduced by the high internal exhaust gas recirculation (EGR),[10] HC and particulate emissions can be a problem. Emissions control to meet strict new and emerging standards will prove difficult. Addressing these problems may increase the weight, cost, and complexity of two-stroke engines, and the outcome of the current intense development efforts is not clear.

Active noise control is another emerging technology of interest. Standard mufflers used to control exhaust noise slightly restrict the flow of the exhaust gases. This increases the pressure in the engine's cylinders during the exhaust stroke and, consequently, increases engine pumping losses, particularly at high engine speed and load. Active noise control can reduce ambient noise by using an electronic device to detect noise (which consists largely of rapid, small variations of ambient air pressure) and to create variations of pressure of equal magnitude and opposite sign that cancel out the original noise. In automotive applications, this approach offers the promise of noise control with little or no restriction on exhaust flow other than that due to the catalytic muffler required for emissions control. However, important uncertainties remain, including the magnitude of the fuel economy advantage as well as the durability, size, and cost of the necessary electronic equipment and generators of

---

[10]EGR involves recycling a portion of the exhaust gases with the fresh air charge.

offsetting noise. Active noise control is used commercially to reduce noise from large air blowers, but it is still in the R&D stage for automotive use.

To explore the bounds of technological feasibility, as well as to achieve favorable publicity, automobile companies develop and present to the public "concept" and "prototype" cars. Such cars may embody extreme styling, new interior concepts, ultra-high speed capability, or other characteristics, such as high fuel economy, low emissions, or extraordinarily safety-conscious design. Data on concept vehicles are useful in establishing achievable boundaries and to confirm and help quantify directional gains. However, the performance of concept and prototype vehicles cannot be directly translated into production vehicles because they do not achieve the variety of objectives that consumers or regulatory agencies require. Typical concept cars and their limitations are discussed in Appendix C.

SUMMARY

- The energy content of automotive fuel is used or dissipated in a variety of ways that are affected by vehicle design and technology and by operating conditions. Each of the pathways of energy use or loss offers the potential for fuel economy improvements. Automobile engineers must seek to optimize the vehicle and its power train to meet the often-conflicting demands of improved fuel economy, customer satisfaction, emission standards, safety, and cost.

- A variety of engine technologies may be applied to achieve improved performance or improved fuel economy. Others offer the promise of improved fuel economy, but at the expense of performance or other attributes that may be valued by consumers.

- Modifying the details of a vehicle's design or technology affects the balance of all of its energy flows. Full consideration of the effects on fuel economy of design or technological changes requires examination of the vehicle as a total system. Modifications may have very complex effects that are difficult to analyze.

- Many proven technologies are available to improve fuel economy. Each technology makes possible small, but important, improvements in fuel economy.

- A number of emerging technologies hold the promise of better fuel economy. At this stage of their development, however, it is impossible to estimate with any accuracy the probability of their success. Nonetheless, it seems reasonably certain that one or more emerging technologies (including some not foreseen in this study) will begin to make significant contributions to fuel economy by 2006.

REFERENCES

Amann, C.A. 1991. Fuel economy factors. Paper presented at the workshop of the Committee on Fuel Economy of Automobiles and Light Trucks, Irvine, Calif., July 8-12.

Associated Press. 1988. Japanese car sets record at 6,509 mpg. Reported in *Knoxville News Sentinel* July 3.

Berger, J.O., M.H. Smith, and R.W. Andrews. 1990. A system for estimating fuel economy potential due to technology improvements. Paper presented at the workshop of the Committee on Fuel Economy of Automobiles and Light Trucks, Irvine, Calif., July 8-12. University of Michigan, Ann Arbor.

DeCicco, J.M. 1991. Cost-effectiveness of fuel economy improvements. Paper presented at the workshop of the Committee on Fuel Economy of Automobiles and Light Trucks, Irvine, Calif., July 8-12. American Council for an Energy-Efficient Economy, Washington, D.C.

Energy and Environmental Analysis, Inc. 1991a. Documentation of Attributes of Technologies to Improve Automotive Fuel Economy. Prepared for Martin Marietta, Energy Systems, Oak Ridge, Tenn. Arlington, Va.

Energy and Environmental Analysis, Inc. 1991b. Fuel economy technology benefits. Presented to the Technology Subgroup, Committee on Fuel Economy of Automobiles and Light Trucks, Detroit, Mich., July 31.

Honda Motor Company, Ltd. 1991. VTEC-E engine. English version of Honda Press Release in Japanese, July 30. Honda North America, Inc., Washington, D.C.

Ledbetter, M., and M. Ross. 1990. Supply curves of conserved energy for automobiles. *Proceedings of the 25th Intersociety Energy Conservation Engineering Conference*. New York: American Institute of Chemical Engineers.

MacCready, P.B., Jr. 1989. Perspectives in transportation. *Journal of the Air Pollution Control Association* November:1428-1429.

Office of Technology Assessment, U.S. Congress. 1991. *Improving Automobile Fuel Economy: New Standards, New Approaches*. Washington, D.C.: U.S. Government Printing Office.

SRI International. 1991. *Potential for Improved Fuel Economy in Passenger Cars and Light Trucks*. Prepared for Motor Vehicle Manufacturers Association. Menlo Park, Calif.

3

SAFETY IMPLICATIONS OF FUEL ECONOMY MEASURES

Of all concerns related to requirements for increasing the fuel economy of vehicles, safety has created the most strident public debate. Some maintain that higher corporate average fuel economy (CAFE) standards will require weight reduction inevitably leading to increased injury and death. Others point to past progress and assert that further substantial gains in fuel economy can be made without sacrificing safety, even if vehicle weight is reduced. Available studies arrive at conclusions that often appear contradictory. Further, because of the differences in the data examined and the analyses conducted, most studies are not directly comparable. In short, although the relationship of fuel economy to safety is of central interest, the issue is surrounded by substantial confusion.

BACKGROUND

Historically, motor vehicle travel in the United States has become steadily safer, a phenomenon observed in every nation as it motorizes. Deaths per vehicle miles traveled (VMT) have continuously decreased from 1930 to the present (see Figure 3-1). This overall trend of improving safety will no doubt continue regardless of U.S. policy on fuel economy.

Many factors have contributed to this trend of reduced fatality rates in the past and are likely to affect future rates. They include support for stricter law enforcement and penalties for driving while intoxicated, increased use of safety belts and increased availability of passive protection, greater urbanization, a lower proportion of very young drivers, reduced motorcycle usage, and continued improvement in the highway environment. Factors operating in the opposite direction include increases in vehicle speeds and increased size and use of large trucks. The aging driving population, with its corresponding increase in crash risk per VMT and vulnerability to injury once a crash occurs, may also be a factor if this age group continues to increase its driving.

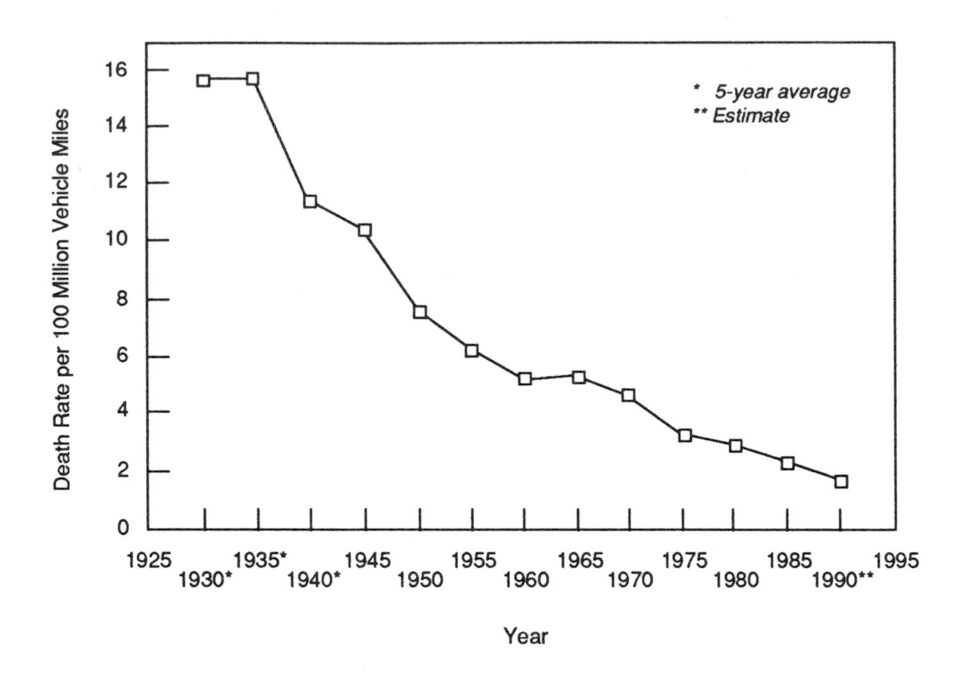

FIGURE 3-1 Death rate per hundred million vehicle miles.

SOURCE: National Safety Council (1990).

Against the background of improving safety there is the question of the impact of changes in the weight and external size of passenger cars and light trucks and in the vehicle fleet mix. This chapter reviews the historical data on relationships between risk of casualty and vehicle size and weight and draws conclusions as to the likely consequences of any future size and weight reductions.

VEHICLE CHARACTERISTICS AND FATALITY RISK

The design characteristics of a vehicle influence its safety in two major ways--by altering the risk of being involved in a collision and by altering the risk of occupant injury if a collision occurs. Because fuel economy measures may lead to changes in the design characteristics of given vehicles or to changes in the vehicle mix, they may affect vehicle safety. In multivehicle collisions, the compatibility of the design of the vehicles involved also affects the risk of occupant injury.

Risk of Crash Involvement

The geometry of a vehicle substantially affects its stability. The propensity of a vehicle to roll over depends on the height of the vehicle's center of gravity and the track width (distance between left and right wheels) of the vehicle. The ratio of these two factors provides an approximate index of rollover stability, although the dynamic characteristics of the suspension and handling also have a major influence. Light trucks and vans are less stable than passenger cars because, for a given track width, their center of gravity is higher. Similarly, because utility vehicles have high ground clearance for off-road operation, they have a greater rollover potential than passenger cars. In fact, the actual rollover rate of utility vehicles is some five times greater than that of passenger cars (Insurance Institute for Highway Safety [IIHS], 1987).[1] Although less marked, smaller passenger cars also have a greater potential for rollover than larger cars.

The risk of crash involvement can also be affected indirectly through policies affecting the characteristics of the vehicle fleet. For example, older cars have higher crash involvement rates per VMT than newer cars so that policies that lead to extension of the life of the vehicle fleet could have adverse safety implications (Stewart and Carroll, 1980). Similarly, as discussed in Chapter 6, improved vehicle fuel economy could encourage increased VMT, thus increasing exposure to risk. On the other hand, high-performance cars (that is, those with short acceleration times) have high occupant fatality rates, and if fuel economy measures diminish the market for or availability of these cars, there would likely be fuel economy and safety benefits (IIHS, 1990). Of course, despite such measures there will still be high risk drivers.

Risk of Injury

Once a collision occurs, the basic physics of the event, coupled with the crashworthiness characteristics of the vehicles and the use of safety belts and passive protection, influence the outcome in terms of the severity of occupant injury.

Fundamentally, two factors influence injury to the occupant. The first is the severity of the crash, usually expressed in terms of the *change* in velocity to the occupant that occurs in the impact. The second is the *distance* over which that change in velocity occurs. A 30-mph change in velocity occurring over 30 feet is hard braking, but is not injurious. A 30-mph change in velocity occurring over 3 feet, the conditions for a severe collision, is normally survivable by a restrained occupant with only moderate and reversible injuries. However, a 30-mph change in velocity occurring over 6 inches would not normally be survivable. These correspond to decelerations of 1, 10, and 60 times the acceleration of gravity, respectively.

Newton's laws of motion dictate that when rigid objects of different weights collide, the velocity change of the lighter object will always exceed that of the heavier object, in proportion to their relative weights. A head-on collision between two cars of comparable structure and equal weight and a closing speed of 60 mph will produce a velocity change for each car of 30 mph. If one car is double the weight of the other,

---

[1]Despite the fact that vans have a much greater rollover potential than passenger cars, they actually experience a very low rate of rollover, presumably because of compensating driver behavior and the different trip characteristics associated with these vehicles.

however, the velocity changes will be 20 and 40 mph, respectively. Hence, in vehicle-to-vehicle collisions, occupants of the heavier vehicle will usually fare better because the change in velocity is less than in the lighter car.

To reduce the deceleration forces on the vehicle occupant, good crash-protection design seeks to provide a long "ride-down" distance (i.e., the distance over which the deceleration occurs) in a vehicle's structure. The greater the change in velocity, the greater the ride-down distance needed. On the whole, the larger the vehicle the greater the ride-down distance available (vans are an exception).

As shown in Table 3-1, which lists occupant fatalities in 1990 for the main crash types (National Highway Traffic Safety Administration [NHTSA], undated c), vehicle-to-vehicle collisions are a large segment of the fatal crash population.[2] Weight and ride-down distance are fundamental to occupant protection, and from the standpoint of individual safety, one is better off in the heaviest possible vehicle that has the largest amount of useful ride-down distance built into its structure. However, as will be discussed subsequently, from an overall societal viewpoint, changes in the weight

TABLE 3-1 Occupant Fatalities in Passenger Vehicles by Vehicle Type, 1990 Fatal Accident Reporting System

		Single Vehicle		
Vehicle Type	Vehicle/Vehicle	Non-Rollover	Rollover	Total
Passenger cars	13,406[a]	5,803	4,816	24,025
Vans[b]	587	259	306	1,152
Pickup trucks[b]	2,333	1,559	2,337	6,229
Utility vehicles[b]	382	194	636	1,212
Total	16,708 (51.2%)	7,815 (24.0%)	8,095 (24.8%)	32,618 (100%)

[a] The number of occupant fatalities in two-car crashes is 4,913. Most of the 13,406 fatalities occur to occupants of passenger cars colliding with a vehicle other than a passenger car.
[b] Small and large.

SOURCE: NHTSA (undated c).

[2]The data in Table 3-1 do not include the approximately 3,200 motorcyclists, 700 medium- and heavy-truck occupants, 6,500 pedestrians, and 850 pedalcyclists who sustained fatal injuries in 1990. Changes in the automotive fleet mix may have some impact on these other fatalities as well.

distribution of vehicles in the fleet produce winners and losers in two-vehicle collisions, the net effect depending on the precise characteristics of the fleet.

Design Compatibility

Vehicle design should seek to optimize protection for the greatest number of people at risk. Consequently, design compatibility among vehicles is an important safety issue.

In a collision between a heavier car and a lighter car, two major factors contribute to design compatibility. The first concerns the dynamic crash characteristics. To achieve greater between-vehicle compatibility and thus protect the occupants of the lighter weight car, the structure of the heavier car should be "softened up" from what is optimal for that car in an impact with a rigid barrier. The second factor concerns geometric characteristics--that is, the cars should "fit together" in a way that minimizes injury. An example of an incompatible fit is that between a pickup truck and a passenger car. The high bumper on the pickup truck leads to greater intrusion into the car than would be the case if the bumper heights of the two were compatible (Hackney et al., 1989). Thus, measures that increase incompatibility through altered vehicle design or a changed mix of vehicles in the fleet would have safety consequences.[3] Such measures could affect occupant safety as well as pedestrian casualties.

SAFETY ISSUES IN IMPROVING FUEL ECONOMY

Three major issues are related to efforts to improve fuel economy. The first, the one receiving the most public attention, is the impact of vehicle downsizing or downweighting on safety.[4] The second is the question of the increased use of light trucks and their safety characteristics. And the third is the question of how much safety can be increased through improved design and technology.

Impact of Downsizing and Downweighting on Safety

With enactment of the Energy Policy and Conservation Act in 1975, the U.S. automobile industry moved quickly to improve fuel economy in passenger cars. Among other changes, the average weight and external size of the U.S. fleet moved toward that in other motorized countries.[5] A variety of studies have concluded that the occupants

---

[3]The pickup truck's high rigid bumpers, often exacerbated by after-market modifications, are particularly hostile to occupants of passenger cars.

[4]Vehicle size and vehicle weight are highly correlated. They are often used interchangeably, and few studies have attempted to measure the effects of each separately.

[5]A corresponding response to the fuel economy crisis was a move to larger, heavier trucks. With the reduction in the maximum speed limit to optimize fuel economy, the trucking industry lobbied successfully for increased cargo capacity (i.e., larger, heavier trucks) in order to remain economically competitive.

of downsized cars during the late 1970s were at increased risk of injury and fatality in comparison with occupants of larger cars. Although more recent passenger cars have marked improvements in safety, overall differences persist between the safety of smaller and larger automobiles.

Some analysts have pointed to these persistent differences and predicted that substantial safety costs will attend further efforts to increase fuel economy. Whether, and to what extent, those claims will hold true cannot be entirely established on the basis of available information. Nonetheless, analyses of historical crash data provide some guidance on the point.

Past Studies

Most past analyses have focused on passenger cars. Between 1975 and 1990 the average wheelbase (distance between front and rear wheels) in the U.S. new passenger-car fleet, including domestic and imported cars, decreased from 110.9 inches to 101.3 inches, and weight declined by about 900 pounds (NHTSA, undated a). Numerous studies have sought to ascertain the safety consequences of these reductions in size and weight.[6]

Several analysts have attempted to quantify the relative safety risk for a vehicle's occupants as a function of vehicle size or weight. Robertson and Baker (1976) found that the smaller cars (wheelbase under 105 inches) were involved in three times as many fatal single-vehicle crashes as the largest cars (wheelbase over 120 inches) and twice as many fatal multiple-vehicle crashes. O'Day and Kaplan (1975) reported that smaller cars (compact and sports models) involved a higher risk of fatality than large cars (standard models) for all driver age groups. The relative risk increased with age; drivers over age 55 experienced a threefold risk compared with drivers aged 16 to 24.

Mela (1975:1) analyzed crash data from New York and North Carolina and concluded that in two-car crashes, "the chance of serious or fatal injury to an unbelted driver decreased by about five percent for each hundred pounds additional weight in his car and increased about two percent for each hundred pounds increase in the weight of the other car in the collision." Jones and Whitfield (1984) analyzed the effects of belt use and car weight on driver injury in one- and two-car crashes. They found that, "each additional thousand pounds of vehicle mass decreases the odds of a driver injury in a crash by about 34 percent when the driver is not restrained. For restrained drivers, this decrease is 25 percent per thousand (p. 51)." Evans and

---

[6]The early studies focused primarily on unrestrained occupants. Currently, however, more than half of front-seat occupants use safety belts. Moreover, for more recent model cars, with their numerous safety improvements, there is less disparity between the safety of larger and smaller vehicles. (See Robertson and Baker, 1976; Campbell, 1974; Campbell and Reinfurt, 1973; Cerrelli, 1984; Chi et al., 1982; Crandall and Graham, 1988; Evans and Wasielewski, 1987; Joksch and Thoren, 1984; Jones and Whitfield, 1984; Mela, 1975; O'Day and Kaplan, 1975; Partyka, 1990; Partyka and Boehly, 1989; Rihlberg et al., 1964; Robertson and Baker, 1976; Stewart and Stutts, 1978; see also IIHS, 1991c). For example, the occupant fatality rates for 1976-1978 model year (MY) cars were almost 3.5 times as high for cars weighing 2,000 pounds or less compared with cars weighing 4,000 to 4,500 pounds. For MY 1986-1988, the ratio was slightly less than 2.5 (Chelimsky, 1991).

Wasielewski (1987:119) reported that, "a driver in a 900 kg [1,980 pounds] car crashing head-on into another 900 kg car is about 2.0 times as likely to be fatally injured or killed as is a driver of an 1800 kg [3,960 pounds] car crashing head-on into another 1800 kg car." More recently, Crandall and Graham (1988:22) reviewed a number of studies of automobile safety and estimated that, "the 500 lb [227 kg], or 14 percent, reduction in the average weight of 1985 cars caused by CAFE standards is associated with a 14 to 27 percent increase in occupant fatality risk."

Partyka and Boehly (1989:75) reported that the shift to lighter cars between 1980 and 1987 would have resulted in a "5.6 percent increase in the number of moderate and greater injuries to drivers if all other factors remained unchanged." However, they pointed out that other factors, such as vehicle design and use of occupant restraints, have not remained unchanged and must be taken into consideration in arriving at conclusions. In a more recent report, Partyka (1990) found that the fatality rate in minicompact cars was twice that in the largest cars.

Recent analyses by NHTSA indicate that the reductions that have occurred in passenger-vehicle size from MY 1970 to 1982 are associated with approximately 2,000 additional occupant fatalities annually (Kahane, 1990; Kahane and Klein, 1991). The additional fatalities identified were in single-vehicle crashes, and approximately two-thirds of them involved rollovers. The primary cause of the rollovers appears to have been the narrower track width of smaller vehicles, a factor that should be resolved independently of fuel economy considerations.

In their analyses of two-vehicle crashes, Evans and Frick (1991c:1) reported that, "if a driver transfers to a car lighter by 1 percent that driver's fatality risk in a two-car crash compared to the risk of the other involved driver, increases by between 2.7 percent and 4.3 percent, the specific value depending on other factors such as model year." They further concluded that, "when other factors are equal, (1) the lighter the vehicle, the less risk to other road users, and (2) the heavier the vehicle, the less risk to its occupants."

In a second study, Evans and Frick (1991b:1) attempted to calculate the relative risk of driver death in two-car collisions. They concluded that,

> when cars of the same mass crash into each other, fatality risk is lower when both cars are heavier. If one of the equal cars is replaced by another lighter by any amount, the fatality risk increase to the driver in the car of reduced mass exceeds the fatality risk reduction for the driver in the unchanged car (that is, net risk increases). Net driver fatality risk (or net fatalities) in a car population increases if any car in the population is replaced by a lighter one or if one population of identical cars is replaced by another population of lighter identical cars.

Other studies have analyzed overall occupant fatality rates in relation to number of registered vehicles (Ford Motor Company, 1991; IIHS, 1987, 1991b) and reported higher fatality rates for smaller cars. The IIHS analyses were based on MY 1984-1988 vehicles involved in crashes during calendar years 1985-1989. Occupant fatality rates for the smaller cars (wheelbase under 95 inches) were approximately twice those of the largest cars (wheelbase over 114 inches). The differential rates were observed for single-vehicle, non-rollover crashes; single-vehicle, rollover crashes; and multivehicle crashes.

While downsizing will typically involve downweighting, the use of appropriate lightweight materials would allow a reduction in vehicle weight without a proportionally similar reduction in size. Because reductions in weight promise fuel economy gains, it is of interest whether the downweighting that is not accompanied by downsizing has a safety penalty. Unfortunately, weight and size are highly correlated in historical data, and it is difficult to disentangle any differential effects. Several analyses have concluded that it is vehicle size, as indicated by wheelbase, rather than weight that is the more important (Ford Motor Company, 1991; IIHS, 1991b; Robertson, 1991). But another study, which attempted specifically to separate the effects of changes in wheelbase from changes in vehicle weight (Evans and Frick, 1991a), concluded that, "mass is the dominant causative factor in the large dependence of driver fatality risk on mass in two-car crashes, with size playing at most a secondary role (p. 1)."[7]

A recent report by the U.S. General Accounting Office (GAO) concluded that automobile weight reductions have *not* led to increased fatalities (Chelimsky, 1991). Certain assumptions and conclusions in the report are subject to question, however. First, the GAO report concluded that, because the overall number of automobile deaths per 100,000 registered vehicles has gone down since the 1970s and passenger cars have become lighter, the weight reductions "have had virtually no effect on total highway fatalities" (p. 2). As pointed out in IIHS's (1991a) response to this report, analyses of the data indicate that the decrease in the overall death rate would have been even greater if the weight reductions had not occurred.

Second, the GAO report assumed that a move toward smaller cars will automatically lead to a decrease in occupant fatalities in two-car crashes because of the "reduced aggressiveness of relatively light cars" (p. 24).[8] Some analyses of two-vehicle crashes have shown, however, that at a given speed a collision between two smaller cars results in greater occupant injury than a similar collision between two larger vehicles (Campbell and Reinfurt, 1973; Evans and Wasielewski, 1987). If that is the case, removing large cars from the vehicle fleet does not necessarily remove the

---

[7]The studies noted are based on somewhat different measures. Evans and Frick, Ford, and IIHS all based their analyses on Fatal Accident Reporting System (FARS) data, but Evans and Frick examined two-car crashes, and Ford and IIHS considered all occupant fatalities in passenger cars. Evans and Frick used the relative risk of driver fatality in two-car collisions, while Ford and IIHS considered the occupant fatality rates per 10,000 registered vehicles. Evans and Frick used very broad ranges for vehicle model years and included much older vehicles; Ford and IIHS considered only more recent models and used much narrower vehicle age ranges. Consequently, while the conclusions reached may appear contradictory, it is entirely possible that each analysis is correct, but that each analysis focuses on a different aspect of the issue. See Appendix D for analyses of hypothetical data sets involving the effects of changes in the weight distribution of cars in the fleet on two-car fatalities and a discussion relevant to the issue.

[8]*Aggressiveness* or *aggressivity* refers to the potential for inflicting damage to the other vehicle and its occupants.

incremental risk to the occupant of the smaller car (Evans and Frick, 1991b). As noted earlier, however, the difference in occupant injury risk between large car-large car collisions and small car-small car collisions has been reduced in more recent model year cars.

Third, the GAO (and other analysts) assumed that drivers in fatal crashes are representative of all drivers. Numerous studies have reported that drivers in fatal crashes are different from other drivers and cannot be considered as representative of the driving population (Baker, 1970; Evans, 1987, 1991; Garretson and Peck, 1982).

The GAO report also pointed out that many of the factors that contributed to the greater risk potential of smaller cars in the past do not necessarily have to persist in the future--for example, the greater propensity to roll over. Thus, while the GAO report recognized the possibility of a trade-off between safety and fuel economy, it concluded that such a trade-off may not be necessary, and that at the very least the predicted safety costs that have been claimed by some are greater than would likely occur.

A recent report by the Office of Technology Assessment (1991) acknowledged that pressures to achieve significant downsizing in a relatively short time period would likely result in a safety cost. However, the report pointed out that if sufficient time is allowed, *and if measures other than downsizing are pursued*, it should be possible to maintain safety while improving fuel economy. The report emphasized the importance of maintaining a vehicle's exterior dimensions and structural integrity, as well as providing sufficient interior volume for occupants to survive the "second crash"--the collision of occupants with the interior of the vehicle. However, the report concluded that reductions in weight need not incur a safety cost, a view not shared by all investigators.

The overall conclusion of previous analyses is that the historical changes in the fleet--downsizing and/or downweighting--have been accompanied by increased risk of occupant injury. Not all studies have found completely consistent relationships between weight change and occupant injury, however. The data do show that the difference between the safety of large and small cars has diminished.

Data Issues

The analysis of safety issues is complicated by a variety of data problems. The use of the FARS data limits analyses to the extreme end of the injury distribution--motor vehicle crashes that result in the death of at least one participant. It is well established that fatal crashes are not representative of the entire spectrum of crashes and do not necessarily provide a valid basis for broad generalizations about crash characteristics.[9]

---

[9]For example, alcohol consumption is much more likely to be a factor in fatal crashes, somewhat less so in injury crashes, and much less so in crashes resulting in property damage only.

Because crashes do not appear in FARS unless they result in at least one fatality, there is a potential bias in the calculation of relative fatality risk in two-car crashes. Based on relative fatality rates derived from FARS, Evans and Frick (1991b) reported that greater homogeneity of the vehicle fleet *increases* fatality risk. This finding could be an artifact of including only crashes in which at least one fatality occurs. To illustrate, in a fatal crash involving a large car and a small car, the probability is greater that the driver of the smaller car will be fatally injured rather than the driver of the larger car. However, in a fatal crash involving two vehicles of equal size (either large or small), the crash severity must be sufficiently great to result in fatal injury to at least one of the drivers in order for the crash to be recorded in FARS. But, if the crash is sufficiently severe to result in fatal injury to one of the drivers, it is more likely to be severe enough to be fatal to the other driver as well, since the two cars, being of nearly equal weight, should experience similar forces.[10] Consequently, if only fatal crashes are considered, greater homogeneity in the size of automobiles will appear to exacerbate the consequences of similarity in size.

Measures of exposure to risk pose additional problems in interpreting available studies. Some of the analyses by Evans and Frick (1991b) are based on the assumption that, for purposes of estimating vehicle exposure to accidents, one can use the distribution of vehicle weights for those vehicles involved in pedestrian fatalities. Yet this measure of exposure requires the assumption that vehicles involved in pedestrian fatalities are representative of the distribution of the exposure of the entire vehicle fleet. What is known about the times and places in which pedestrian fatalities occur, as well as the populations at highest risk for fatal pedestrian injury, raises serious questions about the validity of such an assumption.[11]

While some of the analyses attempt to take some account of driver variables, most do not. Drivers in fatal crashes are not a random sample of drivers in general, and hence the relative risk of fatality attributable to vehicle characteristics *per se* must take into account driver variables and vehicle-use variables.

Individual versus Societal Risk

It is evident that in two-car crashes, other things being equal, the occupants of the heavier car will fare better than the occupants of the lighter car. However, as has been pointed out by a number of investigators, the heavier car, while affording more protection to its occupants, also inflicts more injury on the occupants of the smaller car. Thus, a reduction in the weight of the larger vehicle, while perhaps increasing the risk

[10]Obviously, the fatality risk is also affected by other factors, such as use of occupant restraints, crash geometry, and driver age.

[11]Pedestrian fatalities involve three major groups--children, the elderly, and the alcohol impaired. Using the distribution of car sizes involved in pedestrian fatalities as a measure of exposure of all cars requires two major assumptions. The first is that the exposure of fatally injured pedestrians is representative of the exposure distribution of all car sizes. The second is that once a pedestrian is hit, there is no effect of car size on the probability of fatal injury. Available data regarding the second assumption indicate that smaller cars are less likely to result in fatal injury to a struck pedestrian (e.g., Robertson and Baker, 1976).

for its occupants, carries some benefit for the occupants of the other vehicle. In the case of two-car collisions, it is not clear how these two opposing effects are balanced, at least in the current fleet. Although in the past it has been shown that small car-small car collisions are more injurious than large car-large car collisions (Campbell and Reinfurt, 1973; Evans and Wasielewski, 1987), the difference has diminished over time. Earlier analyses suggested that the safety gains associated with increased vehicle weight were greater than the safety losses associated with the greater aggressivity of heavier cars in collision with other vehicles (Mela, 1975). However, early analyses were based on the assumption that most occupants were unrestrained. Greater use of occupant restraints and improved vehicle design have enhanced occupant safety in more recent model years in ways that could affect the outcome of collisions.

In any event, it is important to distinguish individual risk from societal risk. Estimates of societal risk must take into account the net effects of the safety gains to the occupants of the heavier car and the safety losses that the increased weight imposes on the occupants of the struck car, as well as other road users (e.g., pedestrians, pedalcyclists, and motorcyclists). Whether, and how much, total societal risk is modified by shifts in the mix of vehicle weights is not clear. Based on consideration of two-car collisions using hypothetical estimates of accident variables, Appendix D considers the change in fatalities from various changes in distribution of vehicle sizes in the fleet.[12] The appendix suggests that, in principle, downsizing could increase, decrease, or leave unchanged total deaths and injuries in two-car collisions, depending on the changed size distribution of cars in the fleet. The more aggressive large cars are relative to small and midsize cars, the more beneficial it will be to get large cars off the road. The less crashworthy small cars are relative to larger cars in collisions with cars of equal size, the more beneficial it will be to get small cars off the road. These analyses do not consider other types of crashes--single vehicle into roadside obstacles, passenger car-truck collisions, and collisions of cars and light trucks with pedestrians and cyclists.[13] Moreover, additional analyses are needed, based on more recent model years and actual data, to reflect improved vehicle design and increased use of occupant restraint systems. Presumably, crashes into anything but a rigid barrier would still favor the occupant of the large vehicle, but nonetheless, the shifts in total societal costs that may be anticipated from downsizing passenger vehicles are not clearly established.[14]

Light Trucks

While the reasons for the shift to light trucks are not clear and may have little to do with fuel economy, their increased usage raises concerns about both safety and fuel

---

[12]Appendix D contains a set of sample calculations illustrating this point.

[13] Analysts have suggested that the size of passenger vehicles is related to the probability of fatality even in crashes with large trucks (e.g., Cerrelli, 1984).

[14]The National Highway Traffic Safety Administration declined the committee's request that NHTSA conduct a fuller assessment of this issue. Letter from Administrator J.R. Curry, NHTSA, to R.A. Meserve, chairman, Committee on Fuel Economy of Automobiles and Light Trucks, November 25, 1991.

economy. Light trucks are clearly more likely to roll over than automobiles because their design dictates a higher center of gravity than that for passenger cars.[15] Although the findings from a NHTSA crash-testing program clearly show that recent model light trucks are safer than their earlier counterparts, they still have poorer safety performance than passenger cars.[16]

The rapid expansion of light trucks as a proportion of market share has not been accompanied by a clear understanding of their use. They now represent close to one-third of the market, and according to the industry, about 70 percent are purchased for personal rather than commercial use. In effect, light trucks are being used as a substitute for the passenger car. In the past, however, light trucks have not been subject to the same standards as passenger cars for safety (or fuel economy), nor have the safety standards for light trucks been established in the same manner. Because they have become such a significant portion of the passenger-vehicle fleet, more aggressive efforts to improve the safety and fuel economy of light trucks should be taken. Otherwise, any gains made through improvements in passenger cars may be offset by the increased use of light trucks.

Potential for Enhancing Safety through Design and Technology

Just as estimates of fuel economy improvements are based in part on what is technologically feasible, so too safety should be evaluated with an eye to technological opportunities. For vehicle occupants, future benefits will come from the application of new technologies for crash avoidance and for occupant protection once a collision occurs.

Some safety improvements are already scheduled to be implemented, others are being seriously considered, and still others will occur because of market pressures. Safety modifications are expected to add somewhat to vehicle weight and thus result in some reduction in fuel economy levels. The safety standards already scheduled for implementation include expanded installation of automatic restraints and improved side- and head-impact protection. Additional padding of vehicle interiors may also be required. Antilock brake systems, although not mandated by federal standards, will likely be incorporated into most of the vehicle fleet over the next several years. Based

---

[15]There are anomalies in the data, however. The inherent stability of minivans is much lower than that of passenger cars, which suggests that minivans should have a higher rate of rollover than automobiles, but the actual rollover experience is much lower than that for automobiles. Presumably, the explanation lies in the types of drivers of vans and the types of trips they make (IIHS, 1987).

[16]The National Highway Traffic Safety Administration operates the New Car Assessment Program (NCAP), in which vehicles undergo a standard test crash into a rigid barrier with belted and unbelted dummies. Measurements are made of injury potential, including head injury criteria (HIC) and chest accelerations (chest Gs). The lower the measurements, the less the potential for injury. Light trucks were not included in NCAP testing until MY 1983, and the number of such vehicles tested to date is small (N = 41 as of 1989). Analyses indicate, however, that the measurements for the light-truck fleet are approximately 20 percent higher (less safe) on HIC and 10 percent higher on chest Gs than for passenger cars. When light trucks are further subdivided into light-duty trucks, vans, and multipurpose vehicles, the number within a test category becomes very small. Nevertheless, it is clear that the vans are less safe from the viewpoint of the driver (Hackney et al., 1989).

on the proportion of the new vehicle fleet that would be affected by these changes, there would be an estimated increase of 67 pounds over the fleet's average weight for MY 1990 passenger cars. Similar anticipated changes in light trucks would result in an estimated average weight increase of 133 to 167 pounds above 1991 models. These anticipated weight increases apply to safety improvements only; they do not take into consideration any weight increases associated with technology required for improved emissions control (Bischoff, 1991).

The National Highway Traffic Safety Administration has issued a notice of proposed rulemaking to address the rollover stability of passenger cars and light trucks (Federal Register, 1992). Several options are being discussed, including increasing the stability of the vehicle, improving the crashworthiness of the vehicle if certain levels of stability are not achieved, and providing information on rollover risks to consumers to enable them to make more informed choices. Within each of these categories of action, a variety of options are being considered, as well as possible combinations of the various options. In light of the uncertainty as to the eventual regulation, no attempt has been made here to estimate the weight implications of improvements in rollover stability.

It is also anticipated that increased use of antilock brake systems may reduce the incidence of rollover crashes. Rollovers are usually initiated by a lateral component of movement of the vehicle, and antilock brake systems improve the stability of the vehicle in braking situations, thereby reducing the probability of rollover. As noted, roughly two-thirds of the increased risk of fatality attributed by NHTSA to the downweighting of cars results from rollover (Kahane and Klein, 1991). The incorporation of antilock brake systems into the vehicle fleet might gradually eliminate some of the additional fatalities attributable to downweighting.

As noted, light trucks are not currently subject to the same safety standards as passenger cars. Given the increased use of these vehicles as passenger vehicles (as opposed to cargo carriers), NHTSA is implementing a program of rulemaking that broadly requires the same crash performance for both classes of vehicle. In particular, the extension of occupant crash-protection requirements and the application of mandatory safety belt use laws to occupants of light trucks are likely to diminish the risk to occupants of light trucks and to reduce fatalities (Kee, 1991).

Improvements in vehicle handling, stability, and brakes should lead to significant safety gains for utility vehicles. From the standpoint of societal costs, attention must be given to the lack of design compatibility between light trucks and passenger cars--that is, the geometrical and structural mismatches discussed above. Proposals have been formulated, but the rulemaking process has not yet been initiated on this important safety issue.

Future technologies hold promise for crash avoidance. The use of on-board systems to provide drivers with information not only about vehicle conditions (e.g., underinflation of tires, impending equipment failures) but also about upcoming traffic hazards can enable drivers to avoid otherwise high-risk situations. Intelligent vehicle highway systems include crash avoidance as a major objective.

Once a collision occurs, improved occupant protection can be achieved through advanced restraint systems. Driver- and passenger-side airbags are currently entering the fleet of cars and light trucks. Advanced safety belts with pre-tensioning and load-limiting capabilities will enhance protection, particularly for the more vulnerable segments of the population (e.g., the elderly). Side-impact airbags are under development. The greater acceptance of mandatory safety belt use laws and their extension to all occupants, in front and rear seats, offer significant crash-protection gains.

Better understanding of the behavior of materials subject to impact, through the use of analytic techniques, high-speed computers, and the techniques of computer-aided design, now allow bodyshell designs to be optimized much more efficiently. Thus, technology offers the promise of better compatibility between vehicles of varying weight for varying crash configurations and speeds.

Because safety is becoming increasingly important to the public, it is likely that over the next decade the automotive industry will move beyond the requirements imposed by federal regulations and will offer enhanced safety as a positive marketing attribute. Safety adds value to the product, and the public's growing recognition and appreciation of safety features may well result in greater safety gains than predicted from future regulatory activity alone.

STRATEGIES FOR IMPROVING SAFETY AND FUEL ECONOMY

A number of measures can improve both safety and fuel economy. They can be categorized into three major areas: control of exposure to accidents, crash avoidance, and behavior modification (Trinca et al., 1989).

Exposure Control

To the extent that exposure (i.e., VMT) is reduced, safety and fuel economy can be enhanced.[17] Current examples of measures to reduce exposure include substitution of electronic for physical communication, the provision of low-risk modes of travel (e.g., urban bus networks), and zoning of urban development to diminish commuting distances. Indeed, because discretionary driving often presents the greatest risk, reduction in VMT may offer disproportionate safety benefits (Transportation Research Board, 1984). However, it should be noted that in the United States, efforts to reduce VMT have been singularly unsuccessful--VMT has increased more rapidly than the population of licensed drivers (Highway Statistics, 1987-1990).

[17]The primary purpose of the highway system is mobility, hence exposure. All motor vehicle injury could be prevented if exposure were reduced to zero. Thus, for exposure control to be an effective strategy, it must fulfil the purpose of travel while reducing it.

Crash Prevention

Collisions not only increase the risk of injury, but also contribute to fuel consumption because of the traffic disruptions they cause. Improved highway design and maintenance, as well as improved traffic operations, contribute to safety and fuel economy. While fuel costs increase with road roughness, other operating costs, such as maintenance and repair, increase much more markedly (Wyatt et al., 1979). The potential safety benefits of improved pavement maintenance are reviewed in a summary by Cleveland (1987). Effective traffic management through computerized control of traffic signal networks can optimize traffic flow and minimize speed variance and delays, thereby reducing fuel consumption as well as conflicts and the incidence of collisions.

Improved vehicle design can also contribute to crash avoidance. Daylight running lights, centered high-mounted brake lights, improved truck brakes, and antilock brake systems are examples of such improvements.

Behavior Modification

Driver behavior affects both safety and fuel economy in that fuel-efficient driving is associated with safer performance. An example is the 55-mph National Maximum Speed Limit (NMSL), a measure that was enacted primarily to increase fuel economy. The measure led to significant fuel savings, but the safety benefits were striking as well (National Research Council, 1984).

SAFETY AS A SOCIETAL VALUE

Safety is only one of many societal values that are considered in connection with automotive transportation. Traditionally it has been assumed that "safety does not sell," but recent trends in consumer preference for passenger cars with airbags and for vehicles with good safety records suggest a major shift in the public's concern for safety.[18] Yet the fact remains that the highway transportation system was developed primarily for purposes of mobility, not safety. Decisions must be made regarding the system trade-offs between safety and other values.

Currently, society accepts approximately 45,000 deaths and several million disabling injuries annually as the safety costs for highway transportation in the United States. The public supported the lifting of the 55-mph speed limit on rural interstate highways, although analyses indicated that it would cost several hundred lives annually; subsequent studies have confirmed those predictions (e.g., Baum et al., 1990; National Research Council, 1984; Wagenaar et al., 1990). Right-turn-on-red was promoted in some states as a fuel economy measure, although the evidence suggests it saves little fuel and also results in an increase in pedestrian deaths (Preusser et al., 1982; Zador

---

[18]Ford Motor Company, in its presentation to the committee's Impacts Subgroup on September 16, 1991, provided survey data indicating the rising importance of safety to car buyers between 1980 and 1990.

et al., 1982). The use of motorcycles on public highways is associated with greatly increased crash rates per VMT, as well as extremely high rates of serious injury. The licensing of very young and very old drivers is another area in which safety costs are paid in exchange for mobility.

It may be inevitable that significant increases in fuel economy can occur only with some negative safety consequences. Should that be the case, it does not necessarily follow that increased fuel economy should automatically be rejected. Rather, the choice should be made on the basis of the most complete information available and an evaluation of the relative cost and the relative benefit of the change.

FINDINGS AND CONCLUSIONS

Some maintain that significant increases in fuel economy will require vehicle downweighting which, in turn, will increase injury and death. Others claim that these impacts can be avoided. The fact that both sides are able to make credible arguments attests to the ambiguity of the evidence available and the complications in attempting to forecast the impacts of future changes in vehicle characteristics. Although there are no conclusive answers to the question of whether, and to what extent, safety may be compromised by improvements in fuel economy, the following points may be made:

- Safety, as measured by fatalities per hundred million VMT, has steadily improved since 1930, and it is likely that this general trend will continue. In evaluating the safety consequences of fuel economy measures requiring vehicle modifications, this overall trend must be taken into consideration. Otherwise, safety improvements are likely to be erroneously attributed to changes that are unrelated or even detrimental to safety.

- Numerous studies conducted over the past 30 years show there is a relationship between reduced vehicle size/weight and increased occupant fatality risk, but the difference in safety between large and small cars for recent model years has diminished. Significant improvements have been made in small and large cars, although proportionately greater improvement has been made in small cars, a result that might be anticipated given that small cars had considerably more room for improvement.

- Changes in design or the application of technology, such as a slight increase in track width or the use of antilock brake systems, can differentially reduce the rollover tendency of smaller cars. These changes, however, could increase vehicle weight, with some resulting fuel economy penalty.

- In certain crash types--for example, a collision with a nonrigid fixed object--vehicle weight is a major determinant of the forces experienced by the vehicle occupants. Any major reduction in vehicle weight carries the potential for reduced safety in such collisions. Consequently, any such weight reductions must be accompanied by measures to maintain safety through improved vehicle design or other means.

- While decreased vehicle weight may increase risk in collisions with nonrigid fixed objects, lighter weight vehicles will pose less hazard to occupants of the other vehicle in two-vehicle collisions. Thus, in such collisions, vehicle weight reductions have opposing effects; they decrease the safety of the occupant of the lighter weight vehicle, but increase the safety of occupants of the struck vehicle. It is not clear to what extent these two opposing effects offset each other--that is, the extent to which changes in vehicle weight will have an impact on overall societal costs incurred in two-car collisions. Those studies specifically examining such changes conclude that there is an increase in overall risk with downsizing of the passenger-car fleet.

- Currently scheduled and anticipated vehicle safety modifications--for example, passenger-side airbags, improved side-impact protection, antilock brake systems--will enhance safety at the cost of increased vehicle weight and thus reduce fuel economy. Improved design and the incorporation of new technology, however, can enhance both crash avoidance and crashworthiness potential, while improving fuel efficiency.

- Light trucks have special safety problems, and because they constitute almost one-third of the new vehicle fleet, they warrant careful scrutiny. Moreover, light trucks are particularly aggressive to passenger cars in passenger car-light truck collisions. Some measures to reduce such aggressivity--for example, lowering bumpers--should not affect fuel economy.

- Reducing exposure, improving crash avoidance, and increasing driving efficiency improve both safety and fuel economy. These measures should be pursued independently of technological measures to increase fuel economy.

- There is likely to be a safety cost if downweighting is used to improve fuel economy (all else being equal), although the available information is insufficient to make a specific estimation of the impact.

- Safety is an important consideration in fuel economy deliberations, but it must be considered in relation to other important societal values that are affected by improved fuel economy. Concern for safety alone should not be allowed to paralyze the debate on the desirability of enhancing the fuel economy of the fleet.

- Because of the importance of the safety issue, coupled with the equivocal interpretation of existing and projected data, a comprehensive study of the effects of vehicle weight and size on safety should be conducted and should examine the full range of crash severity and crash types. While such a study will not provide definitive answers regarding the future safety performance of the vehicle fleet, it should be valuable in clarifying the dimensions of the safety aspects of measures to improve fuel economy in automobiles and light trucks.

REFERENCES

Baker, S.P. 1970. *Characteristics of Fatally Injured Drivers*. Final Report. School of Hygiene and Public Health. Baltimore, Md.: John Hopkins University Press.

Baum, H.M., J.K. Wells, and A.K. Lund. 1990. Motor vehicle crash fatalities in the second year of 65 mph speed limits. *Journal of Safety Research* 21:1-8.

Bischoff, D. 1991. *Weight Impacts of Safety Standards*. Washington, D.C.: National Highway Traffic Safety Administration.

Campbell, B.J. 1974. *Driver Injury in Automobile Accidents Involving Certain Car Models: An Update*. Highway Safety Research Center. Chapel Hill: University of North Carolina.

Campbell, B.J., and D.W. Reinfurt. 1973. *Relationship Between Driver Crash Injury and Passenger Car Weight*. Highway Safety Research Center. Chapel Hill: University of North Carolina.

Cerrelli, E.C. 1984. *Risk of Fatal Injury in Vehicles of Different Size*. Washington, D.C.: National Highway Traffic Safety Administration.

Chelimsky, E. 1991. *Automobile Weight and Safety*. Statement before the Subcommittee on Consumers, Committee on Commerce, Science, and Transportation, U.S. Senate. GAO/T-PEMD-91-2. Washington, D.C.: U.S. General Accounting Office.

Chi, G.Y., D.W. Reinfurt, C.V. Britton, A.Y. Leung, and W.C. Fischer. 1982. *Driver Injury in Accidents Involving Certain Vehicle Subgroups Classified by Make/Model and Model Year, Make/Model, and by Market Size*. Highway Safety Research Center. Chapel Hill: University of North Carolina.

Cleveland, D.E. 1987. *Effect of Resurfacing on Highway Safety*. State of the Art Report 6, *Relationship Between Safety and Key Highway Features, A Synthesis of Prior Research*. Washington, D.C.: Transportation Research Board.

Crandall, R.W., and J.D. Graham. 1988. *The Effect of Fuel Economy Standards on Automobile Safety*. The Brookings Institution and Harvard School of Public Health. NEIPRAC Working Paper Series, No. 9. Cambridge, Mass.: New England Injury Prevention Center.

Evans, L. 1987. Fatal and severe crash involvement versus driver age and sex. American Association of Automotive Medicine (AAAM), *31st Annual Conference Proceedings*. Des Plaines, Ill.: AAAM.

Evans, L. 1991. *Traffic Safety and the Driver.* New York: Van Nostrand Reinhold.

Evans, L., and M.C. Frick. 1991a. *Car Size or Car Mass--Which Has Greater Influence on Fatality Risk?* GMR-7462. Warren, Mich.: General Motors Research Laboratories.

Evans, L., and M.C. Frick. 1991b. *Driver Fatality Risk in Two-Car Crashes--Dependence on Masses of Driven and Striking Car.* GMR-7420. Warren, Mich.: General Motors Research Laboratories.

Evans, L., and M.C. Frick. 1991c. *Mass Ratio and Relative Driver Fatality Risk in Two-Vehicle Crashes.* GMR-7419. Warren, Mich.: General Motors Research Laboratories.

Evans, L., and P. Wasielewski. 1987. Serious or fatal driver injury rate versus car mass in head-on crashes between cars of similar mass. *Accident Analysis and Prevention* 19:119-131.

Federal Register. 1992. Vol. 57, No. 2, January 3. National Highway Traffic Safety Administration, 49 CRF, Part 571 (Advanced Notice of Proposed Rulemaking): 242-252.

Ford Motor Company. 1991. *Relationship Among Car Wheelbase, Car Weight and FARS Rates.* Dearborn, Mich.

Garretson, M., and R.C. Peck. 1982. Factors associated with fatal accident involvement among California drivers. *Journal of Safety Research* 13(4):141-156.

Hackney, J.R, W.T. Hollowell, and D.S. Cohen. 1989. *Analysis of Frontal Crash Safety Performance of Passenger Cars, Light Trucks and Vans and An Outline of Future Research Requirements.* 89-2A-0-001. Washington, D.C.: National Highway Traffic Safety Administration.

Highway Statistics. 1987-1990. Washington, D.C.: Federal Highway Administration.

Insurance Institute for Highway Safety (IIHS). 1987. Small passenger vehicles a problem. *Status Report* 22(2). Arlington, Va.

Insurance Institute for Highway Safety (IIHS). 1990. Where is the safety in the fuel economy debate? *Status Report* 25(8). Arlington, Va.

Insurance Institute for Highway Safety (IIHS). 1991a. General Accounting Office "just plain wrong" in conclusions about automobile weight reductions and motor vehicle deaths. News Release. Arlington, Va.

Insurance Institute for Highway Safety (IIHS). 1991b. Occupant deaths per 10,000 registered cars, 1984-88 models in calendar years 1985-89. Various figures. Arlington, Va.

Insurance Institute for Highway Safety (IIHS). 1991c. Wishful thinking: Comments on the report "The Safe Road to Fuel Economy" by Donald Friedman and Keith D. Friedman of MCR Technology, Inc., and Clarence M. Ditlow and Douglas C. Nelson. Center for Auto Safety, Arlington, Va.

Joksch, H.C., and S. Thoren. 1984. *Car Size and Occupant Fatality Risk, Adjusted for Differences in Drivers and Driving Conditions*. Hartford, Conn.: The Center for the Environment and Man.

Jones, I., and R. Whitfield. 1984. The effects of restraint use and mass in "downsized" cars. *Advances in Belt Restraint Systems: Design, Performance, and Usage*. SAE 840199. Washington, D.C.: Insurance Institute for Highway Safety.

Kahane, C.J. 1990. *Effect of Car Size on Frequency and Severity of Rollover Crashes*. Washington, D.C.: National Highway Traffic Safety Administration.

Kahane, C., and T. Klein. 1991. *Effect of Car Size on Fatality and Injury Risk*. Washington, D.C.: National Highway Traffic Safety Administration.

Kee, O. 1991. *Basis for Current Regulations on Fuel Economy and Safety of Light Trucks and Vans*. Washington, D.C.: National Highway Traffic Safety Administration.

Mela, D.F. 1975. *A Statistical Relationship Between Car Weight and Injuries*. Washington, D.C.: National Highway Traffic Safety Administration.

National Highway Traffic Safety Administration (NHTSA). Undated a. Changes in average passenger car size, by wheel base. Chart for years 1970-1990. Washington, D.C.

National Highway Traffic Safety Administration (NHTSA). Undated b. Changes in average weight of passenger cars. Chart for years 1977-1990. Washington, D.C.

National Highway Traffic Safety Administration (NHTSA). Undated c. Occupant fatalities by passenger cars, vans, light trucks, and utility vehicles, by crash type--single and multi-vehicle, rollover and non-rollover. Washington, D.C.

National Research Council. 1984. *55: A Decade of Experience*. Transportation Research Board Special Report 204. Washington, D.C.: National Academy Press.

National Safety Council. 1990. *Accident Facts*. Chicago, Ill.

O'Day, J., and R. Kaplan. 1975. *How Much Safer Are You in a Large Car?* Presented at the Society of Automotive Engineers' Automotive Engineering Congress and Exposition, February 24-28. SAE 750116. Ann Arbor, Mich.: Highway Safety Research Institute.

Office of Technology Assessment, U.S. Congress. 1991. *Improving Automobile Fuel Economy: New Standards, New Approaches*. Washington, D.C.: U.S. Government Printing Office.

Partyka, S.C. 1990. Differences in reporting car weight between fatality and registration data files. *Accident Analysis and Prevention* 22:161-166.

Partyka, S.C., and W.A. Boehly. 1989. Passenger car weight and injury severity in single vehicle nonrollover crashes. *Proceedings of the Twelfth International Technical Conference on Experimental Safety Vehicles.* ESV Report No. 89-2B-0-005. Washington, D.C.: National Highway Traffic Safety Administration.

Preusser, D.F., W.A. Leaf, K.B. DeBartolo, R.D. Blomberg, and M.M. Levy. 1982. The effect of right-turn-on-red on pedestrian and bicyclist accidents. *Journal of Safety Research* 13:45-55.

Rihlberg, J.K., E.A. Narragon, and B.J. Campbell. 1964. *Automotive Crash Injury in Relation to Car Size*. No. VJ-1823-R11. Buffalo, N.Y.: Automotive Crash Injury Research of Cornell University.

Robertson, L.S. 1991. How to save fuel and reduce injuries in automobiles. *Journal of Trauma* 31:107-109.

Robertson, L.S., and S.P. Baker. 1976. Motor vehicle sizes in 1440 fatal crashes. *Accident Analysis and Prevention* 8:167-175.

Stewart, J.R., and C.L. Carroll. 1980. *Annual Mileage Comparisons and Accident and Injury Rates by Make, Model*. Washington, D.C.: National Highway Traffic Safety Administration.

Stewart, J.R., and J.C. Stutts. 1978. *A Categorical Analysis of the Relationship Between Vehicle Weight and Driver Injury in Automobile Accidents*. Highway Safety Research Center. Chapel Hill: University of North Carolina.

Trinca, G., B.J. Campbell, F.A. Haight, I. Johnston, P. Knight, G.M. Mackay, A.J. McLean, and E. Petrucelli. 1989. *Reducing Traffic Injury--A Global Challenge*. Melbourne: Royal Australia College of Surgeons.

Wagenaar, A.C., F.M. Streff, and R.H. Schultz. 1990. Effects of the 65 mph speed limit on injury morbidity and mortality. *Accident Analysis and Prevention* 22:571-585.

Wyatt, R.J., R. Harrison, B.K. Moser, and L.A.P. de Quadros. 1979. The effect of road design and maintenance on vehicle operation costs--Field research in Brazil. In *Low Volume Roads: Second International Conference*. Transportation Research Record 702. Washington, D.C.: Transportation Research Board.

Zador, P., J. Moshman, and L. Marcus. 1982. Adoption of right turn on red: Effects on crashes at signalized intersections. *Accident Analysis and Prevention* 14:219-234.

4

ENVIRONMENTAL ISSUES

In 1990 Congress enacted and the President signed the Clean Air Act amendments imposing, among other elements, new federal regulations on automotive emissions (P.L. 101-549). The law is intended to reduce ground-level ozone and carbon monoxide (CO) in areas in the United States that do not meet ambient air quality standards. As will be discussed, this statute and its associated regulations will limit potential fuel economy improvements by adding weight to vehicles and impeding the use of existing and emerging technologies for improving fuel economy. This chapter examines briefly the nature of automotive emissions, trends in automotive emissions control and air quality, current and prospective clean air standards, the need to reduce emissions further, and the direct and indirect effects of emissions controls on fuel economy.

THE NATURE OF AUTOMOTIVE EMISSIONS

Air pollution from motor vehicles arises from evaporative emissions from the fueling systems of volatile organic compounds (VOCs) that include hydrocarbons (HCs) and from postcombustion chemical compounds that leave the engine through the exhaust (tailpipe) system and the crankcase. In engines using unleaded gasoline, the compounds in the exhaust are typically HCs, CO, and oxides of nitrogen (NO_x); in diesel engines, they include particulates that are related to smoke; and in engines using alternative fuels such as methanol, they include such VOC compounds as formaldehyde. Exhaust emissions are a function of engine operation--for example, compression ratio, spark timing, air/fuel ratio (A/F), and postengine treatment for purposes of control. Hydrocarbons in the exhaust are incompletely burned or unburned fuel and oil. Carbon monoxide is formed in the combustion process and is always present in small quantities in the exhaust regardless of the air/fuel ratio. The greater proportion of fuel there is in the air/fuel mixture, the more CO is produced. Oxides of nitrogen are formed during the combustion process, increase with peak combustion temperature, and are also a function of the air/fuel ratio.

Automotive pollutants, directly and indirectly, have adverse health effects and their discharge into the atmosphere has been subject to regulatory control for over two decades.[1] Exhaust emissions can be limited by a variety of means. Exhaust gases that escape past the piston rings into the crankcase are drawn back into the engine using a positive crankcase ventilation system, and the unburned HCs are combusted. Emissions released through the exhaust pipe are controlled in virtually all vehicles today by three-way catalytic converters in the exhaust system and by electronic controls on gasoline-powered engines. The introduction in 1978 of the three-way catalyst marked a major stride in emissions control technology because it enabled the limitation of NO_x, HC, and CO emissions to levels in compliance with current standards.[2]

Hydrocarbon emissions from the fuel system that occur while the car is in operation or while parked are controlled using a carbon canister that adsorbs the vapors. Such controls have been in use since 1975. Hydrocarbon losses during refueling, when vapor is displaced from the fuel tank by the entering liquid, can be controlled either by returning vapor from the vehicle to the service station tank (referred to as Stage 2 control) or by using a larger carbon canister on the vehicle that traps the fuel vapors. Stage 2 controls on fuel pumps in service stations are already used in some jurisdictions (e.g., California and Washington, D.C.); on-board control is contemplated, but it is currently not required.

Carbon dioxide (CO_2), a greenhouse gas, is a common and necessary by-product of any engine that burns carbon-based fuels, but its potential effect on global climate is a cause for increasing international concern.[3] Recent studies have addressed the issue of possible global warming from projected increases in the concentrations of greenhouse gases--CO_2, methane, nitrous oxide (N_2O), and chlorofluorocarbons (CFCs)--in the atmosphere (National Academy of Sciences, 1991; OTA, 1991a).[4] The

---

[1]Carbon monoxide displaces oxygen from hemoglobin in the blood stream and is considered a causative or contributing factor in a number of health problems. Photochemical oxidants (smog) that include ground-level ozone are formed in the atmosphere in the photochemical reaction of VOCs and NO_x emitted from the tailpipes of vehicles. Oxides of nitrogen are considered hazardous, and ozone is a debilitating irritant, especially of the respiratory system. It has been estimated that several hundred million incidents per year of respiratory symptoms might be avoided if all areas of the United States were in compliance with federal ground-level ozone standards, at a savings of $0.5 to $4 billion annually (Office of Technology Assessment [OTA], 1989). Animal toxicologic and human epidemiologic studies suggest that ozone plays a role in the initiation of respiratory disease processes. In addition, short-term health effects are associated with ozone exposure, including impairment of lung function. Many health professionals believe that lifetime exposure to high levels of ozone may result in premature aging of the lungs, but no reliable estimate of the effect on life expectancy is available.

[2]In this system, platinum and rhodium are used in the front part of the converter to reduce NO_x, and palladium and platinum are used in the rear part to oxidize HCs and CO. It is called a three-way catalyst because it controls HC, CO, and NO_x.

[3]The United States is the world's largest contributor of CO_2 emissions from anthropogenic sources--one-quarter of the world's total (National Academy of Sciences, 1991).

[4]The effects of N_2O are distinct from the effects of NO_x, which by common usage refers to nitric oxide (NO) and nitrogen dioxide (NO_2). Nitrous oxide is a greenhouse gas; NO_x is active in the photochemical reaction that produces ozone. Greenhouse gas emissions from human activities represent the following percentages of total

U.S. fleet of light trucks and automobiles accounts for about one-fifth of the total U.S. CO_2 emissions, and any improvements in fuel economy, assuming an equal number of miles traveled, would lead to concomitant reductions in CO_2 emissions (National Research Council, 1990; OTA, 1991). If, for example, the fuel consumption of the vehicle fleet decreased by 10 percent, total U.S. emissions of CO_2 would decline by about 2 percent.[5]

Chlorofluorocarbons are used as the working fluid in automotive air-conditioning systems. Approximately 0.6 million metric tons of CFCs are released each year. They are potent greenhouse gases and will eventually be eliminated. General Motors is planning the complete removal of CFCs from its new-car fleet by 1995, and other manufacturers are expected to adopt similar schedules.[6]

AUTOMOTIVE EMISSIONS CONTROL AND AIR QUALITY: A BRIEF HISTORY

Concern with air quality, including ground-level ozone, increased during the 1960s and led to federal and state motor vehicle standards for emissions of HCs, CO, and NO_x. Enabling legislation was passed in California in 1965 and at the federal level in the Air Quality Act of 1967 and the Clean Air Act amendments of 1970, 1971, and 1977 (P.L. 91-604, P.L. 93-319, P.L. 95-95; Johnson, 1988; Rau and Wooten, 1980; Shiller, 1990). The federal legislation included authorization for the U.S. Environmental Protection Agency (EPA) to regulate mobile and stationary sources of pollution, establish National Ambient Air Quality Standards (NAAQS), require states to submit implementation plans to meet standards, and issue compliance orders or bring suits against offenders.

By 1988, emissions per vehicle mile traveled (VMT) for new cars and light trucks had been decreased by roughly 90 percent from the uncontrolled level. Total VMT, however, increased 2.3 percent annually during the 1970s and 1980s, thereby offsetting some of the improvement. Consequently, vehicles remain a significant source of pollution in most major urban areas. In 1987, cars and light trucks accounted for 30 percent of the U.S. emissions of VOCs and NO_x and 60 percent of CO emissions (Atkinson et al., 1990). Today, the ambient standard for CO emissions is being exceeded in about 60 urban areas, and approximately 115 million Americans live in 98

---

emissions on a CO_2-equivalent basis: CO_2, 66.3; methane, 20.4; CFCs, 9.7; and N_20, 3.6 (National Academy of Sciences, 1991).

[5]Such a reduction would not necessarily affect automobile emissions of HC, CO, and NO_x because the regulations are expressed in grams per mile driven and are not related to the amount of fuel consumed. Increased fuel economy could nonetheless reduce total HC emissions because, if less gasoline is needed, the systemwide emissions of HC would be reduced (DeLuchi et al., 1991).

[6]Limitations on the most effective refrigerants, CFCs, have already been endorsed for objectives unrelated to global warming, namely, protection of stratospheric ozone levels. Action to accelerate the scheduled phase-out of the CFCs is now under consideration by the Congress (New York Times, February 7, 1992).

areas (in 1990) that do not meet federal ozone standards (ozone nonattainment areas) (Atkinson et al., 1990; National Research Council, 1991). Further, emissions of VOCs are expected to increase in the 1990s if additional controls are not imposed. The problem is exacerbated by deterioration and modification (i.e., tampering) of vehicle emissions control devices during use.[7]

The percentage of older cars on the road is also increasing, which reduces the rate of reduction in automotive emissions. The proportion of the car fleet that is 12 years old or older rose from 11.9 percent in 1980 to 20.8 percent in 1990, and the proportion in the light-truck fleet grew from 18.4 to 27.6 percent (Motor Vehicle Manufacturers Association, 1991). Older vehicles account for a disproportionate share of emissions, even though they are driven fewer miles per year than new cars. For example, although pre-1981 vintage vehicles accounted for only about 35 percent of VMT in 1988, they accounted for about 70 percent of HC emissions, 75 percent of CO_2 emissions, and 68 percent of NO_x emissions.[8]

Congressional and public support for efforts to improve air quality has been strong and led to passage of the Clean Air Act amendments of 1990. These amendments address ambient air quality levels and, in particular, the ozone nonattainment problem through more stringent controls on automotive emissions, requirements for alternative clean fuels, controls on industrial facilities, and other measures. California's regulations will be more stringent than the federal ones, and under the Clean Air Act amendments they can be adopted by states that are not in compliance with ambient air quality standards.

STANDARDS IN THE 1990 CLEAN AIR ACT AMENDMENTS

Federal Standards

The Clean Air Act amendments of 1990 established stringent tailpipe emissions standards for nonmethane hydrocarbons (NMHC),[9] CO, NO_x, and particulates for passenger cars and light trucks of 6,000 pounds gross vehicle weight (GVW) rating or

---

[7]In 1988, results of a field test in Los Angeles that measured pollutant concentrations in air flowing out of the Van Nuys road tunnel suggested that vehicle emissions were two to four times higher than levels predicted by the vehicle-emissions inventory models (Ingalls et al., 1989; National Research Council, 1991).

[8]The disparity in emissions increases with vehicle age both because older cars may have had more lenient emissions requirements at the time of manufacture and the control equipment deteriorates over time. Automobiles that are at least 15 years old accounted for less than 8 percent of VMT in 1988, yet they accounted for about 23 percent of HC emissions, 22 percent of CO_2 emissions, and 18 percent of NO_x emissions (Atkinson et al., 1990). On the other hand, longer vehicle life, while it adds to both direct fuel use and emissions, reduces waste materials requiring disposal and reduces energy consumption and other impacts associated with car production.

[9]Hydrocarbon standards exclude methane, which does not contribute to the formation of ozone or urban smog. This exclusion becomes relevant in considering the problem of global climate change because methane is a greenhouse gas.

less (sce Table 4-1) (P.L. 101-549). The first set of requirements (Tier I) are to be passenger cars and light trucks of 6,000 pounds gross vehicle weight (GVW) rating or less (see Table 4-1) (P.L. 101-549). The first set of requirements (Tier I) are to be phased in beginning with model year (MY) 1994, and 100 percent compliance is to be achieved by MY 1996. By the end of 1999, the EPA will determine the need, cost, and feasibility of additional standards (Tier II) for vehicles produced for MY 2004 and thereafter.

The Clean Air Act amendments also require the EPA administrator to issue regulations to reduce evaporative HC emissions from all gasoline-fueled vehicles and to limit the volatility of gasoline. In addition, the EPA administrator is to issue standards requiring on-board diagnostics to detect emissions-system malfunctions and impose controls on vehicle-refueling emissions by light-duty vehicles.

California Standards

California, as noted, has established emissions standards that are more stringent than the federal standards. California's standards impose emissions levels for five categories of vehicles: (1) conventional vehicles (CVs); (2) transitional low-emission vehicles (TLEVs); (3) low-emission vehicles (LEVs); (4) ultra-low emission vehicles (ULEVs); and (5) zero-emission vehicles (ZEV's) (see Table 4-2). A gradual phase-in of these increasingly "clean" automobiles is planned as manufacturers produce the

TABLE 4-1 Passenger-Car Emissions Standards (grams per mile), Clean Air Act Amendments of 1990

	Gasoline Engines			Diesel Engines	
Standard	NMHC	CO	NO_x	PM	NO_x
Current (1991)	0.41T	3.4	1.00	0.20	1.0
Tier I	0.25 (0.31)	3.4 (4.2)	0.4 (0.6)	0.08 (0.10)	1.0 (1.25)
Tier II	(0.125)	(1.7)	(0.2)	(0.08)	(0.2)

NOTES: NMHC = nonmethane hydrocarbons; T = total hydrocarbons; CO = carbon monoxide; NO_x = oxides of nitrogen; PM = particulate matter. Standards are for 5 years/50,000 miles or (10 years/100,000 miles) and for up to 3,750 pounds loaded vehicle weight. Tier I standards must be achieved by MY 1996; Tier II, if imposed, would apply to MY 2004 and beyond.

TABLE 4-2 California's Passenger-Car Emissions Standards (grams per mile)

	Gasoline Engines			Diesel Engines
Vehicle Class	NMOG	CO	NO_x	PM
93 base	0.25 (0.31)	3.4 (4.2)	0.4	(0.08)
TLEV	0.125 (0.166)	3.4 (4.2)	0.4 (0.6)	(0.08)
LEV	0.075 (0.090)	3.4 (4.2)	0.2 (0.3)	(0.08)
ULEV	0.040 (0.055)	1.7 (2.1)	0.2 (0.3)	(0.04)
ZEV*	0	0	0	0

NOTES: NMOG = nonmethane organic gases; CO = carbon monoxide; NO_x = oxides of nitrogen; PM = particulate matter; TLEV = transitional low-emission vehicle; LEV = low-emission vehicle; ULEV = ultra LEV; ZEV = zero-emission vehicle. Standards are for 5 years/50,000 miles or (10 years/100,000 miles) and for up to 3,750 pounds loaded vehicle weight. For 1993 base, NMOG = nonmethane hydrocarbons only.

*In 1998, 2 percent of manufacturers' sales must be ZEVs and in 2003, 10 percent.

vehicles in four phases during the 1990s and on to 2003 (Chang et al., 1991). To meet California's goal to reduce nonmethane organic gases (NMOG), the manufacturers will have to achieve a decreasing fleet-average standard for these emissions using a combination of the vehicles identified above.[10]

Under the provisions of the Clean Air Act amendments, the 31 other nonattainment states may adopt California's standards, and a number of states intend to do so.[11] The states that have taken action or are considering it represent about half of the new-car market in the United States. Given the problems associated with marketing cars with different standards in different states, all automobiles may be

[10]Nonmethane organic gases are equivalent to nonmethane hydrocarbons. For passenger cars and light-duty trucks whose loaded vehicle weight (LVW; curb weight plus 300 pounds) is less than or equal to 3,750 pounds, the standard will evolve from 0.390 grams per mile (gpm) for NMOG in MY 1992 (compliance with the fleet average would be required in MY 1994) to 0.062 gpm in MY 2003. For light-duty trucks greater than or equal to 3,751 pounds LVW, the standard will move from 0.5 gpm in MY 1992 to 0.093 gpm in MY 2003. For NO_x, emissions levels of 0.2 gpm are prescribed for LEVs and ULEVs.

[11]Delaware, the District of Columbia, Maine, Maryland, Massachusetts, New Hampshire, New Jersey, New York, Pennsylvania, and Virginia have signed an agreement to adopt the California standards (New York Times, October 30, 1991).

designed to meet the more stringent emissions standards if they become widely accepted, even though air quality conditions differ from region to region. This could mean that California's standards would control planning and decision making on future emissions control systems, not the federal requirements.

Emission Standards and Technology Development

An attractive technology for improving fuel economy involves the introduction of excess air (and/or exhaust gas recycling) during combustion--the lean-burn approach. For NO_x control under lean-burn conditions using excess air, the current three-way catalyst is ineffective. A completely new catalyst is needed for NO_x control under these conditions, and as discussed below, the outlook for such a development is uncertain.

The tiered structure of the regulations may inhibit technology development, however. Manufacturers have no incentive to introduce technologies that meet intermediate levels of control because of the possibility that regulations could quickly make the new technologies obsolete. In addition, the new requirements for increased durability of emissions systems (from 50,000 to 100,000 miles) may also have a detrimental effect on the introduction of new technologies because it will be more difficult to obtain the required operating experience with the new technology to be certain of adequate performance. In addition, this requirement will require changes in current designs in order to exceed the standards during initial operation (e.g., by adding increased amounts of catalyst) to offset deteriorating performance. These changes will add cost and may impose increases in vehicle weight.

In sum, control of NO_x and HC emission to meet Tier II and California LEV and ULEV levels presents many substantial challenges to automotive manufacturers.[12] As discussed below, compliance with the standards will make it difficult to introduce more fuel efficient vehicles.

EMISSIONS CONTROL AND FUEL ECONOMY

Current gasoline-engine catalytic systems approximate optimality as a result of 15 years of operating experience. Modern engines operate in a relatively efficient combustion region and ensure that 98 to 99 percent of the carbon in the fuel is converted to CO_2 in the engine. In short, essentially all of the energy in the fuel is released by efficient combustion.[13]

---

[12]Control of exhaust CO was not addressed because, in general, controls for exhaust HC also control for CO.

[13]Some small percentage of the fuel is burned late in the cycle because of trapping in engine crevices, in the oil, and so forth. Late burning results in inefficient conversion of fuel energy into work.

Reducing emissions further will not improve fuel economy.[14] In fact, emissions control reduce fuel economy by increasing vehicle weight[15] and limiting the opportunities for improved fuel economy. Automotive manufacturers indicated in presentations to the committee (see Appendix F) that the 1996 federal Tier I emissions standards will increase weight by 45 to 66 pounds and reduce fuel economy by 1.5 to 1.8 percent. California's standards are expected to reduce fuel economy by an additional 1.5 to as much as about 4 percent.[16]

Control of Oxides of Nitrogen

Automotive manufacturers will emphasize improving the performance of the three-way catalyst to meet future NO_x standards until a new catalyst is available for control of NO_x under lean-burn conditions (i.e., air/fuel ratio greater than stoichiometric).[17] The Tier II NO_x standard will be particularly difficult to meet; it is equivalent to about 95 percent control from precontrol levels.[18] New catalyst systems will also be required to meet California's HC standard for LEVs and ULEVs, which will probably entail fuel economy penalties.

The current three-way catalyst system can be modified to reduce NO_x emissions further by using larger amounts of noble metals (particularly rhodium) in the catalytic converter,[19] controlling the air-fuel ratio more precisely, introducing an electrically heated catalyst (or light-off catalyst with secondary air) to improve effectiveness during cold starts, and recycling exhaust gases to lower the peak temperature in the cylinders. In addition, further reduction of sulphur in gasoline would reduce the catalyst's light-off temperature and would allow the use of palladium and cerium in the catalytic system (Monroe et al., 1990). Even so, it is not certain that the level of NO_x (0.2 gpm)

---

[14]The mass of emissions from automobiles is directly proportional to VMT. Any actions that reduce VMT will generally reduce total emissions of HC, CO, NO_x, and CO_2.

[15]Emissions-control equipment on 1991 models accounts for about 25 pounds in cars and 35 pounds in light trucks (Kelly Brown, Ford Motor Company, personal communication, December 1991).

[16]In presentations to the committee, Honda estimated that Tier I standards will reduce economy by 1.5 percent and increase vehicle weight by 45 to 66 pounds and that California's standards will reduce fuel economy by an additional 1.5 percent. Ford estimated that the Tier I standards will reduce fuel economy by 0.5 miles per gallon (mpg) and increase weight by 20 pounds. Nissan estimated that the Tier I standards will reduce fuel economy by 1.8 percent and that California's LEV standard will lead to a 5.6 percent decline in fuel economy.

[17]In stoichiometric operation, the air/fuel ratio is set so that the mixture incudes precisely the amount of oxygen needed for complete combustion of the fuel. Under lean-burn conditions, excess oxygen (air) is added.

[18]In addition, the new standards require 100,000 miles durability for emission-control equipment, compared with 50,000 miles currently.

[19]One study indicates that roughly three times as much rhodium is required to achieve 0.4 gpm NO_x as to reach 0.7 gpm (Sierra Research, 1988). The increased cost for the extra rhodium would be about $39 per car, assuming the cost of rhodium is $3,000 per ounce. The modification corresponds to a cost of $1,900 per ton of NO_x removed, based on amortizing the $39 incremental catalyst cost over 10 years at 10 percent, 0.3 gpm incremental removal of NO_x and 10,000 miles driven per vehicle-year.

that is required by the new standards can be met, and in any event, the modifications would add weight and reduce fuel economy. In addition, no manufacturer is optimistic about meeting the Tier II NO_x standard (0.2 gpm) in gasoline-powered systems, even with a new NO_x catalyst.

It is easier in general to meet the NO_x standards with lighter weight vehicles than with heavier ones.[20] Some improvement in NO_x control may be achieved in heavier vehicles by using oversized engines with internal exhaust-gas recycling (EGR) as a means of reducing the peak cylinder temperature. Such a strategy, combined with the three-way catalyst, EGR, and electrical heating of the catalyst, may make it possible for the emissions of heavier vehicles to approach the Tier II standard. But such an approach will add weight, reduce efficiency, and increase cost. It is possible that the lowest levels of NO_x control cannot be met with the current mix of vehicles.

The Technical Challenge: Lean NO_x Catalyst

Substantial efforts are under way to develop NO_x control technologies for lean-burn systems (gasoline or diesel). Researchers are exploring new catalytic compositions, in particular, zeolites containing copper and other metals (Hamada et al., 1990; Iwamoto, 1990; Iwamoto et al., 1991; Li and Hall, 1991; Montreuil and Gandhi, 1991; Sato et al., 1991).

Systems that lower NO_x emissions in the exhaust from the engine (so-called engine-out emissions) by 40 to 60 percent are available in laboratories. The catalysts are temperature sensitive, however, and under exhaust conditions typical of high-speed operation, they tend to deactivate, presumably because of the agglomeration of the active metal and deterioration of the zeolite. Thus, the systems require increased control of exhaust temperature. Moreover, the volume of catalyst required is large, which increases heat and back pressure in the exhaust system, which in turn lowers fuel economy. The new catalysts are also sensitive to oxygen and sulfur. Because of the oxygen sensitivity, these catalysts may be of limited value for diesel exhaust systems and ultra-lean gasoline engines (e.g., A/F of 22 to 25).[21] The sulfur sensitivity could require a substantial reduction of the sulfur content of fuel--for example, to 30 parts per million or less. Moreover, it may be necessary to enrich the HC concentrations in the exhaust to enhance reduction activity, which makes it more difficult to meet the HC

[20]This arises from the fact that, all else being equal, the volume of exhaust gas varies directly with the weight of the vehicle. Because the emissions standard is given in grams per mile, the permissible exhaust NO_x concentration that satisfies a given standard is larger in the lighter vehicle.

[21]Because diesel exhaust is generally at a lower temperature than gasoline-engine exhaust, the temperature control problem associated with the new catalysts could be ameliorated in diesel applications. However, there remains a problem with the particulates in diesel exhaust; they are likely to accumulate on the NO_x catalyst, thereby lowering its effectiveness.

exhaust standards without an additional oxidation catalyst downstream of the NO_x catalyst.[22]

One of the automotive manufacturers with an aggressive R&D program in lean NO_x catalysts expressed optimism about developing a durable catalyst that will achieve a 50 percent reduction of the NO_x in the engine-out exhaust. Based on information provided to the committee by some of the manufacturers, however, more than 75 percent catalytic reduction of the engine-out NO_x is required to meet the Tier II standard of 0.2 gpm NO_x. It appears that development of lean NO_x catalytic control will be possible, but in view of the problems being encountered, it may not be possible to meet the lowest level of NO_x (0.2 gpm) that has been promulgated in all vehicles. This could mean that the heaviest vehicles would not qualify and, under the worst case, would have to be discontinued. Because NO_x control may be critical to achieving satisfactory levels of urban ozone, appropriate alternatives should be considered (National Research Council, 1991).

Alternative NO_x Control Strategies

Cars and light-duty trucks emitted about 6 million tons of NO_x in 1985. About 2.8 million tons were emitted from other transportation sources, 6.8 million from large utility boilers, 1.8 million from other large boilers, and 1.8 million from small stationary sources (EPA, 1989). Although cars and light trucks are significant contributors of NO_x emissions, they are hardly the only culprits.

Reductions in NO_x emissions from stationary sources have not kept pace with those from mobile sources over the past two decades. In fact, much of the reduction from large stationary sources has been offset by growth in emissions from small stationary facilities (Atkinson et al., 1990). In new large plants, such as utility boilers, up to 90 percent reduction of the NO_x emissions is possible using existing technology, and somewhat lower percentage reductions are possible in retrofitted applications and in smaller facilities.[23] Existing standards, however, do not appear to require such significant reductions from stationary sources. In light of the difficulty of achieving further stringent NO_x reductions in automobiles, perhaps reduction from stationary

[22] Use of lean systems will also result in the oxidation of NO to NO_2 in the exhaust system. As a result, at a given level of NO_x, the NO_2 concentration in the exhaust will be increased, perhaps increasing the ground-level concentration of NO_2 (Li and Hall, 1991). Although NO_2 is more toxic than NO, this change is not expected to be a problem because ambient levels of NO_2 currently are satisfactory. However, higher NO_2 levels at a given level of NO_x could increase the ground-level concentration of ozone because NO_2 is the photochemically active component, not NO.

[23] Control is accomplished using catalytic or noncatalytic chemical treatment (Boer et al., 1990; Epperly et al., 1988; Sedman and Ellison, 1986). Costs compare quite favorably with those for incremental NO_x removal in automobile sources (OTA, 1989), without taking into account any fuel economy penalty. Even higher levels of control in stationary sources should be possible using combinations of technology (combustion modification and chemical treatment).

sources should be contemplated as an offset for relaxing the new controls on light-duty vehicles.[24]

In areas with the greatest ozone problems, NO_x from stationary sources is a smaller percentage of total NO_x emissions than the national average (OTA, 1989). As a result, a given percentage relaxation of the automotive NO_x standard would require a disproportionately larger increase in NO_x control from stationary sources to achieve the same impact on air quality in those areas. Before Tier II standards are adopted nationwide, a detailed assessment should be made to determine whether it is more practical on a region-by-region basis to achieve higher levels of NO_x control from stationary sources than to achieve 95 percent control for mobile sources of NO_x.

Control of Hydrocarbons

Reductions in VOCs of roughly 30 to 50 percent are thought to be necessary to bring most ozone nonattainment areas into compliance (Atkinson et al., 1990). Cars and light-duty vehicles account for about 30 percent of VOC emissions. Thus, the need for further control of automotive exhaust HCs is understandable.

Heated Catalysts

As noted earlier, the 0.075 gpm NMOG standard for LEVs in California is likely to result in added cost and vehicle weight and a consequent reduction in fuel economy, and the ULEV standard is even more stringent. An electrically heated catalyst may be required to meet California's HC standard of 0.075 gpm for LEVs, according to tests by the California Air Resources Board (1990). One manufacturer told the committee that such a catalyst would add up to 97 pounds of weight to a vehicle and reduce fuel economy by 3 percent. A study by the Automotive Consulting Group (1991) indicates, however, that an electrically heated catalyst would add 40 pounds, reduce fuel economy by about 0.9 mpg, and cost $822 to $1,045 per car. Efforts are under way to develop cost-effective approaches for controlling HC emissions.[25] A recent announcement by Ford's research and engineering center at Dunton, Essex indicates, however, that a much simpler system that avoids electrical heating has good potential to be a strong competitor to other known methods for catalyst heating (Griffiths, 1991).[26] It appears that less expensive systems will be developed, but California's standards will continue to present significant challenges.

---

[24]The committee appreciates that such a strategy might require the modification of existing legislation.

[25]For example, the catalyst might be heated initially by briefly igniting a mixture of fuel and air in an "afterburner" slightly upstream of the catalyst. Fuel could be supplied by calibrating the engine to run with excess fuel. Air could be supplied by an electrically driven pump. Additionally, by suitable control of fuel to individual cylinders, a mixture with both excess air and excess fuel could be supplied to the catalyst. However, this approach would not be effective until the catalyst reached the initial light-off temperature.

[26]It is not clear whether this development is directly applicable to cars meeting U.S. standards. The impact of this development on the manufacturers' ability to meet California's NMOG standard of 0.04 gpm for ULEVs is also unclear.

Gasoline Volatility

The EPA administrator is expected to promulgate controls on gasoline volatility during the five-month summer period. Compared with 1985 levels, this would reduce VOCs in 1994 by 12 percent in ozone nonattainment areas and by 14 percent in attainment areas (OTA, 1989). In 1994, the VOC reduction would be 1.3 million tons in nonattainment areas and 3.3 million tons nationwide.

In 1992, the EPA administrator is also expected to regulate gasoline vapor losses during refueling. These vapor losses are the last significant source of vehicle emissions that is largely uncontrolled. As noted, there are two alternative means for controlling vapor losses, control on the vehicle (on-board) and control at the service station (Stage 2 controls).[27] In 1999, on-board control is projected to eliminate about 180,000 tons of VOCs annually in ozone nonattainment areas and 370,000 tons nationwide, a 1.6 percent reduction from the level in 1985 (OTA, 1989). Stage 2 systems could have a greater impact--up to 10 percent greater control if installed in all service stations (Sierra Research, 1988).

Stage 2 controls are effective immediately regardless of the age of the car, whereas on-board controls become effective only with the turnover of the vehicle fleet. In addition, the weight of the on-board system (about 5 pounds for cars) is avoided, and fuel economy is thus improved incrementally.[28] Also, Stage 2 controls provide the flexibility to use the system only where it is needed. Two recent studies indicate that Stage 2 control is at least as cost-effective (dollars per ton of HC emissions eliminated) as on-board control and may be significantly more cost-effective (OTA, 1989; Sierra Research, 1988). In the committee's view, recovery of refueling vapors at the service station should be given careful attention as an alternative to on-board control.

Reformulated Gasolines

Also in 1992, the EPA is expected to issue regulations calling for cleaner, reformulated gasolines in nine ozone nonattainment areas. Regulation of benzene, other aromatic HCs, heavy metals, and oxygenated HCs is expected. In anticipation of these regulations, the automotive manufacturers and the oil industry formed a cooperative program, the Auto/Oil Air Quality Improvement Research Program, to assess the effects of changes in gasoline composition on emissions. Results thus far indicate that HC emissions can be reduced in MY 1983-1985 and MY 1989 vehicles by 5 to 9 percent by including 15 percent of methyl tertiary butyl ether (MTBE, an

---

[27]Control is achieved by use of a special nozzle on the gasoline pump that allows the capture of HCs that would otherwise be expelled during refueling.

[28]The National Highway Traffic Safety Administration (1991) has also expressed concern that on-board vehicle vapor recovery systems may not be safe.

oxygenated HC) in gasoline.[29] Use of MTBE will lower fuel economy (mpg) as a result of the lower volumetric heat content per gallon, but it would have no effect on the use of crude oil, assuming MTBE is made from natural gas. Changes in concentrations of aromatics, olefins, and heavier HCs have had variable effects on emissions. Also, one of the automotive manufacturers told the committee that, individually, these latter changes would lower fuel economy by up to 2.8 percent, largely because of the decrease in volumetric heat content of the fuel. These reductions are not additive, and estimates of the combined effects of changes in fuel composition that might be made are not yet available. There might be some offsetting savings in the operation of refineries making gasoline, but the matter requires further study.

Sulfur in Gasoline

The Auto/Oil program has also shown that reducing the sulfur content of gasoline from 466 to 49 parts per million reduces HC emissions by 16 percent, CO by 13 percent, and NO_x by 9 percent in 10 MY 1989 cars (Auto/Oil Air Quality Improvement Research Program, 1991; Benson et al., 1991). Emissions prior to catalytic treatment were not affected by sulfur, so the change is due to improved performance of the three-way catalyst. Other data show that sulfur inhibits the conversion of NO_x to ammonia, and presumably lower sulfur fuels would lead to higher ammonia emissions. Additional work is planned to define better the effect of sulfur concentration and compound type on emissions.

Alternative Fuels

Evaporative and exhaust HC emissions would also be reduced with the use of alternative fuels. Both methanol and compressed natural gas (CNG) are expected to be used in fleets of 10 or more vehicles to control emissions further. In addition,

---

[29]The following results (all expressed in percent) have been reported by the Auto/Oil group (Hochhauser et al., 1991):

	Model Year					
	1983-1985			1989		
Change	THC	CO	NO_x	THC	CO	NO_x
Aromatics: from 45 to 20%	+13.6	- 2.5	-11.0	- 6.5	-13.3	+2.0
MTBE: from 0 to 15%	- 9.1	-14.1	- 1.3	- 5.5	-11.1	+1.4
Olefins: from 20 to 5%	+ 5.7	- 1.8	- 6.7	+ 5.8	+ 1.5	-6.1
T90: from 360 to 280°F*	- 5.8	+13.6	+ 2.2	-21.7	+ 1.2	+4.9

*T90 refers to the temperature at which 90 percent of the fuel boils. It is related to the amount of heavier HCs in the fuel.

methanol-gasoline mixtures are likely to be used more generally.[30] It appears feasible to modify light-duty vehicles to use 100 percent methanol and CNG and maintain performance that is generally comparable to that of gasoline-fueled vehicles.

Control of Stationary Sources

It is estimated that about 52 percent of VOCs emanate from stationary sources and about 48 percent from mobile sources in ozone nonattainment areas. Further, the percentages are roughly constant regardless of the ozone level (OTA, 1989). Organic solvent evaporation, surface coatings, and the petroleum and gas industries are the major stationary sources of VOCs. Clearly, opportunities to control these stationary sources should be carefully considered as an offset to more stringent HC standards for automobiles.

Impact of Emissions Standards on Light Trucks

In the past, emission standards have affected light trucks in much the same way as they have affected cars, and this trend is expected to continue. As noted earlier, MY 1991 cars have an average of 25 pounds of emissions-control hardware, compared with 35 pounds on light-duty trucks; and meeting the 1996 emissions standards will add another 20 pounds to cars and about 25 pounds to light trucks. These numbers do not include the hardware required to meet Tier II or California LEV standards.

The future emissions standards for light trucks weighing up to 3,750 pounds (LVW)[31] (27 percent of MY 1991 sales) are projected to be essentially the same as those for cars, considering both the federal (Tier I) and California (LEV) requirements (EPA, 1991). As a consequence, it will be possible to use emissions-control hardware that is similar to that on cars, as long as the performance characteristics important in light trucks can be maintained. Lean-burn operation with gasoline is less likely to be used on vehicles that carry substantial payloads because the higher load factor means that lean burn can only be used for a smaller portion of the operating time.

Emissions standards for heavier trucks, which constituted over 70 percent of light-truck sales in 1991, have been adjusted to higher levels in consideration of vehicle

---

[30]In the 1990s, it is estimated that the use of methanol-gasoline blends will effectively reduce the rates of evaporative and exhaust VOC emissions by 30 percent, compared with vehicles meeting current standards and operating on low-volatility fuel (vapor pressure of 9 pounds per square inch) (OTA, 1989). Over the longer term, there is the potential to reduce VOC emission rates by up to 90 percent, but improved engine design and catalysts that control formaldehyde emissions are required.

The EPA has estimated that exhaust emissions from a dual-fueled vehicle while operating on CNG would effectively be reduced by 40 percent and evaporative emissions by 100 percent. However, the latter does not include the emission of methane due to fueling and distribution losses. Methane losses during pipeline transmission and distribution to end users are estimated to be 0.5 to 4 percent of the energy transported (Ho and Renner, 1991). Thus, the widespread use of CNG to fuel vehicles could have an adverse impact on global climate change if a significant fraction of vehicles were fueled with CNG.

[31]LVW is loaded vehicle weight, which is curb weight plus 300 pounds.

weight and payload. For example, the base federal NO_x standard for the MY 1994 is 0.4 gpm for trucks weighing up to 3,750 pounds (LVW) and 0.7 for those up to 5,750 pounds, both less than 6,000 pounds GVWR.[32] Similarly, the California LEV standards for NO_x are 0.2 and 0.4 gpm, respectively. The committee met several times with the automotive manufacturers, and at no time were truck emissions standards raised as presenting problems that were different from those for cars.

It is likely that the Tier I exhaust emissions standards for trucks will be met largely by improving the performance of the three-way catalyst, as with cars. Lean-burn operation is a possibility at the lower end of the weight spectrum of light trucks, but as noted, it is less likely at the heavier end.

Information presented elsewhere in this chapter raises doubts regarding the probability of meeting future emissions standards with the diesel engine. The prospects may be somewhat better in larger trucks (6,000 to 8,500 pounds GVWR) because of the somewhat less stringent NO_x standards. In addition, heavier and more expensive control systems might be feasible on the larger, more expensive trucks. However, in general the emissions standards will increase the percentage of trucks that use gasoline engines because of the NO_x and particulate emissions characteristics of diesel engines.[33] Manufacturers told the committee that a 25 to 50 percent reduction in NO_x from the current 1 gpm requirement might be achieved in diesels using combustion modification. They thought, however, that higher levels of control would require use of methanol fuel instead of diesel and even then were not optimistic about meeting the 0.2 gpm Tier II standard. Tier I standards call for a reduction of particulates from the current 0.2 to 0.08 gpm. While this appears possible, further development is required.

Surveillance of Existing Vehicles

One alternative to increasingly stringent emissions standards for new vehicles is surveillance and servicing of existing vehicles. The EPA has estimated that emissions reductions from a comprehensive inspection and maintenance program could exceed 30 percent for HC and 50 percent for CO and would be roughly 5 to 10 percent for NO_x (Sierra Research, 1988). The emissions from a car with an ineffective catalyst can be equivalent roughly to those from 10 to 20 well-functioning cars or more. In short, the gains from improved surveillance and maintenance are large.[34]

A number of states currently monitor emissions from vehicles by measuring pollutants when the engine is idling. These programs will expand; the EPA has

---

[32]GVWR is the maximum weight for which a vehicle is designed.

[33]There are three approaches to controlling diesel particulates: higher fuel-injection pressure, oxidation catalysts, and particulate traps. These approaches can be combined and, in fact, the most cost-effective system currently utilizes higher pressure injection and an oxidation catalyst.

[34]In addition to encouraging the repair of malfunctioning control systems, a surveillance program would discourage tampering with control systems. The retirement of noncomplying cars would also improve fuel economy because many would be replaced with new cars with higher fuel economy.

established guidelines that require state inspection and maintenance programs to achieve at least 25 percent reductions of HCs and CO. On-board diagnostic equipment is expected on MY 1994 cars to detect malfunctions that can cause vehicles to exceed emissions requirements, and this equipment offers new opportunities for ensuring adequate emissions performance by the fleet. The new on-board diagnostic equipment will substantially change the type of surveillance program that is required, however.

INDIRECT IMPACTS

Improvements in vehicle fuel economy will have important indirect environmental impacts. For example, replacing the cast iron and steel components of vehicles with lighter weight materials (e.g., aluminum, plastics, or composites) may reduce fuel consumption but would generate a different set of environmental impacts, as well as result in different kinds of indirect energy consumption. Depending on the energy intensities of the substitute materials, and the manufacturing processes employed, those indirect costs could offset to varying degrees the improvements in vehicle fuel economy. Although the committee has not done an extensive analysis of these effects, it appears that the production of a large automobile requires the energy equivalent of two to three years of gasoline consumption, with a corresponding set of emissions to the environment. (The energy consumed may be of forms that are not petroleum based.) The recycling of discarded vehicles to a greater extent than is the case today could reduce the net energy required to manufacture new vehicles and would conserve resources.[35] However, such programs could limit the use of plastics and composites because of the special difficulties they pose in the recycling system.

Fuel economy improvements reduce total HC emissions, considering emissions from the production, transportation, refining, and marketing of fuel (DeLuchi et al., 1991). But greenhouse gas emissions from the production of substitute materials, such as aluminum, could substantially offset decreases of those emissions achieved through improved fuel economy, although this would strongly depend on the energy source used for the production of the materials. Additionally, such indirect impacts should be important considerations in evaluating alternative-fuel or electric vehicles (Amann, 1990; Electric Power Research Institute, 1991).

CONCLUSIONS

- The 1996 federal Tier I emissions standards can be met with a loss in fuel economy of roughly 0.5 mpg.

- The Tier II and California standard of 0.2 gpm of NO_x may not be feasible for the current mix of automobiles and light trucks, and it is likely to be most difficult to meet for the largest vehicles. In addition, these standards are likely to foreclose

[35]In Europe, the automobile companies have initiated a more intensive recycling program. A similar program is under consideration by the U.S. companies (see Automotive News, September 23, 1991).

broad application of promising current and emerging technologies for improving fuel economy, such as lean-burn engines, particularly for larger vehicles. Although the Tier II standard is yet to be adopted, the California standard is being implemented, and it is being seriously considered by states that collectively account for half of U.S. sales of cars and light trucks. In view of the difficulty of meeting these NO_x standards, other approaches, such as increasing control of NO_x from stationary sources, should be considered.

- It appears possible that a lean NO_x catalytic system achieving roughly a 50 percent reduction in NO_x with the required durability will be developed. Such a system might make it possible to meet Tier II and California LEV standards in the smallest cars and trucks, but the system is unlikely to achieve the control needed in the heaviest light-duty vehicles without an additional, substantial breakthrough in catalyst technology.

- Diesel-powered vehicles are unlikely to meet the Tier II and California LEV standards for NO_x and particulates. Thus, the substantial fuel economy benefit of the diesel engine is unlikely to be available when these standards take effect.

- Lighter weight vehicles have the potential for lower NO_x emissions and higher fuel economy and thus offer health advantages in improved air quality and lower CO_2 emissions. As a result, trends toward lower emissions levels and higher fuel economy will favor lighter vehicles.

- In view of the importance of HC emissions in the formation of ground-level ozone, service station control of fueling emissions (Stage 2) seems to promise distinct advantages over on-board recovery. Once installed, such a system would be effective for all vehicles, not just new ones. In addition, the Stage 2 system can be applied only in areas where it is needed. In view of these benefits, service station control of refueling emissions should be given further consideration by the EPA.

- Emissions of HC, CO, and NO_x from existing vehicles probably can be significantly reduced by lowering the sulfur content of gasoline. In addition, lowering sulfur content would reduce the light-off temperature and increase the flexibility to use other catalytic metals in new vehicles. The EPA should consider implementation of further sulfur reduction as part of its program on reformulated gasoline.

- Emissions of HC and CO from the current fleet could generally be reduced by adding MTBE to gasoline. This would reduce fuel economy in terms of miles per gallon, but it would not increase the use of petroleum, assuming the MTBE is made from natural gas. Other changes in gasoline composition have mixed effects.

- Surveillance and enforcement of emissions standards in the existing fleet may be an attractive alternative to increasingly stringent controls on new vehicles.

REFERENCES

Amann, C.A. 1990. Technical options for energy conservation and controlling environmental impact in highway vehicles. *Energy and Environment in the 21st Century*. Proceedings of conference held at Massachusetts Institute of Technology, Cambridge, Mass., March 26-28. Cambridge, Mass.: Energy Laboratory.

Atkinson, D., A. Cristofaro, and J. Kolb. 1990. Role of the automobile in urban air pollution. *Energy and Environment in the 21st Century*. Proceedings of conference held at Massachusetts Institute of Technology, Cambridge, Mass., March 26-28. Cambridge, Mass.: Energy Laboratory.

Auto/Oil Air Quality Improvement Research Program. 1991. *Effects of Fuel Sulfur Levels on Mass Exhaust Emissions*. Technical Bulletin No. 2. Dearborn, Mich.: Ford Motor Company.

The Automotive Consulting Group, Inc. 1991. *The Electrically Heated Catalyst: A "Systems" Cost Analysis*. Ann Arbor, Mich.

Benson, J.D., V. Burns, R.A. Gorse, A.M. Hochhauser, W.J. Koehl, L.J. Painter, and R.M. Reuter. 1991. *Effects of Gasoline Sulfur Level on Mass Exhaust Emissions-Auto/Oil Air Quality Improvement Research Program*. SAE Paper 912323. Warrendale, Pa.: Society of Automotive Engineers.

Boer, F.P., L.L. Hegedus, T.R. Gouker, and K.P. Zak. 1990. Controlling power plant *NO_x emissions*. *ChemTech* 20(5):312.

California Air Resources Board. 1990. *Low-Emission Vehicle Technical Support Document*. Los Angeles.

Chang, T.T., R.H. Hammerle, S.M. Japor, and T.J. Saleem. 1991. Alternative transportation fuels and air quality. *Environmental Science and Technology* 25(7):1190-1197.

Congressional Quarterly. 1990. Clean Air Act amendments. *For the Record* 24 (November) 24:3934-3963.

DeLuchi, M.A. 1990. State-of-the-art assessment of emissions of greenhouse gases from the use of fossil and nonfossil fuels, with emphasis on alternative transportation fuels. University of California, Davis.

DeLuchi, M.A. 1991. Greenhouse gas emissions from the use of transportation fuels and electricity. Argonne National Laboratory, Ill. Photocopy.

DeLuchi, M.A., Q. Wang, and D.L. Greene. 1991. Motor vehicle fuel economy, the forgotten HC control strategy? Oak Ridge National Laboratory, Tenn. Photocopy.

Electric Power Research Institute. 1991. They're new! They're clean! They're electric! *EPRI Journal* April/May:4-15.

Epperly, W.R., J.E. Hofmann, J.H. O'Leary, and J.C. Sullivan. 1988. The NOx OUT(R) process for reduction of nitrogen oxides. Paper presented at the 81st A.P.C.A. Annual Meeting and Exhibition, Dallas, June 23.

Griffiths, J. 1991. Ford puts out a cleaner cat. *Financial Times* November 28.

Hamada, H., Y. Kintaichi, M. Sasaki, T. Ito, and M. Tabata. 1990. Highly selective reduction of nitrogen oxides with hydrocarbons over H-form zeolite catalysts in oxygen-rich atmospheres. *Applied Catalysis* 64:L1-L4.

Ho, S.P., and T.A. Renner. 1991. *Global Warming Impact of Gasoline vs. Alternative Transportation Fuels*. SAE Paper 901489. Warrendale, Pa.: Society of Automotive Engineers.

Hochhauser, A.M., J.D. Benson, V. Burns, R.A. Gorse, W.J. Koehl, L.J. Painter, B.H. Rippon, R.M. Reuter, and J.A. Rutherford. 1991. *The Effect of Aromatics, MTBE, Olefins, and T90 on Mass Exhaust Emissions from Current and Older Vehicles*. SAE Paper 912322. Warrendale, Pa.: Society of Automotive Engineers.

Ingalls, M.N., L.R. Smith, and R. E. Kirksey. 1989. *Measurement of On-Road Vehicle Emission Factors in the California South Coast Air Basin*. Vol. 1: *Regulated Emissions*. San Antonio, Texas: Southwest Research Institute.

Iwamoto, M. 1990. Catalytic decomposition of nitrogen monoxide. *Studies in Surface Science and Catalysis* 54:121-143.

Iwamoto, M., H. Yahiro, K. Tanda, N. Mizuno, and Y. Mine. 1991. Removal of nitrogen monoxide through a novel catalytic process. 1. Decomposition on excessively copper ion exchanged ZSM-5 zeolites. *Journal of Physical Chemistry* 93:3727-3730.

Johnson, J.H. 1988. Automotive emissions. In *Air Pollution, the Automobile, and Public Health*. The Health Effects Institute. Washington, D.C.: National Academy Press.

Khazzoom, J.D. 1991. A model of the auto manufacturers' target emission rate for new passenger vehicles. San Jose State University, Calif. Photocopy.

Li, Y., and W.K. Hall. 1991. Catalytic decomposition of nitric oxide over Cu-zeolites. *Journal of Catalysis* 129:202-215.

Monroe, D.R., M.H. Krueger, and D.J. Upton. 1990. The effect of sulfur on three-way catalysts. GMR-7135. General Motors Research Laboratories, Warren, Mich.

Montreuil, C.N., and H.S. Gandhi. 1991. Lean NO_x catalyst. Ford Motor Company. Dearborn, Mich.

Motor Vehicle Manufacturers Association. 1991. *Facts & Figures '91.* Washington, D.C.

National Academy of Sciences. 1991. *Policy Implications of Greenhouse Warming.* Washington, D.C.: National Academy Press.

National Highway Traffic Safety Administration (NHTSA). 1991. *An Assessment of the Safety of Onboard Refueling Vapor Recovery Systems.* Prepared by Enforcement, NHTSA. Washington, D.C.

National Research Council. 1991. *Rethinking the Ozone Problem in Urban and Regional Air Pollution.* Washington, D.C.: National Academy Press.

Office of Technology Assessment (OTA), U.S. Congress. 1989. *Catching Our Breath: Next Steps for Reducing Urban Ozone.* Washington, D.C.: U.S. Government Printing Office.

Office of Technology Assessment (OTA), U.S. Congress. 1991a. *Changing by Degrees: Steps to Reduce Greenhouse Gases.* Washington, D.C.: U.S. Government Printing Office.

Office of Technology Assessment (OTA), U.S. Congress. 1991b. *Improving Automobile Fuel Economy: New Standards New Approaches.* Washington, D.C.: U.S. Government Printing Office

Quarles, J., and W.H. Lewis, Jr. 1990. *The NEW Clean Air Act, A Guide to the Clean Air Program as Amended in 1990.* Washington, D.C.: Morgan, Lewis and Bockius.

Rau, J.G., and D.C. Wooten. 1980. *Environmental Impact Analysis Handbook.* New York: McGraw-Hill.

Sato, S., Y. Yu-u, H. Yahiro, N. Mizuno, and M. Iwamoto. 1991. Cu-ZSM-5 zeolite as highly active catalyst for removal of nitrogen monoxide from emission of diesel engines. *Applied Catalysis* 70:L1-L5.

Sedman, C.B., and W. Ellison. 1986. German FGD/$DeNO_x$ experience. *Proceedings of the Third Annual Pittsburgh Coal Conference.* Sponsored by the University of Pittsburgh and the U.S. Department of Energy's Pittsburgh Energy Technology Center. Pittsburgh, Pa.

Shiller, J.W. 1990. The automobile and the atmosphere. In *Energy: Production, Consumption, and Consequences*, J.L. Helm, ed. National Academy of Engineering. Washington, D.C.: National Academy Press.

Sierra Research, Inc. 1988. *The Feasibility and Costs of More Stringent Mobile Source Emission Controls*. Prepared for the Office of Technology Assessment, U.S. Congress. Sacramento, Calif.

U.S. Environmental Protection Agency (EPA). 1989. *The 1985 NAPAP Emissions Inventory (version 2): Development of the Annual Data and Modelers' Tapes*. Office of Research and Development. EPA-600/7-89-012a. Washington, D.C.

U.S. Environmental Protection Agency (EPA). 1991. *Implementation Strategy for the Clean Air Act Amendments of 1990*. Office of Air and Radiation. Washington, D.C.

5

IMPACTS ON THE AUTOMOTIVE INDUSTRY

The U.S. automotive industry is facing a difficult if not unprecedented period of competition and capital spending in its efforts to compete with Japanese automakers and to meet pending government regulations on emissions control and safety. These burdens are falling on an industry trying to cope with massive losses due to the 1990-1991 recession and the battle for market share.

Fuel economy has not been a major competitive issue in the marketplace since 1981. Relatively low gasoline prices have allowed consumers to focus instead on vehicle prices, performance, comfort, and style. But firms that can provide all of those characteristics, plus superior fuel economy, should have an advantage over those that do not. Whatever the future fuel economy standards, U.S. automakers must confront the fact that the Japanese appear to have targeted improved fuel economy as an area deserving particular emphasis.[1]

Although properly designed fuel economy standards would not necessarily constitute a competitive disadvantage to U.S. automotive companies (see Chapter 9), new fuel economy standards that are extremely costly to implement or that greatly distort the normal product cycle of the industry would place an enormous financial burden on domestic automakers. Hence, the impacts on the industry are central to the discussion of new fuel economy targets. This chapter explores those impacts.

---

[1]It is difficult to separate fact from rhetoric on this matter. The Japanese are targeting fuel economy in Japan (U.S. Department of Commerce and Motor Equipment Manufacturers Association, 1990) under the threat of tougher U.S. corporate average fuel economy (CAFE) laws and the possibility that Japan will adopt its own version of CAFE standards, and because the price of gas is nearly $4 a gallon there and the Japanese government has been critical of declining fuel economy. Nonetheless, it must be noted that in general, on a class-by-class basis, U.S. cars currently have better fuel economy than similar Japanese cars. In its presentation on November 4, 1991, to the Standards and Regulations Subgroup of the committee (see Appendix F), General Motors noted that for MY 1990 cars, if the sales mix of its vehicles had been the same (by size class and transmission types) as its principal Japanese competitors, its CAFE rating would have been as follows: GM with a Honda mix, 31.6 vs. Honda's 30.8 mpg; GM with a Nissan mix, 29.4 vs. 28.4 mpg for Nissan; GM with a Toyota mix, 30.3 vs. 30.6 mpg for Toyota. General Motor's actual CAFE rating for MY 1990 was 27.1 mpg.

FINANCIAL PERFORMANCE

The domestic manufacturers will continue to confront serious financial burdens, wholly apart from any changes in fuel economy regulations. The domestic automotive industry is mature and highly cyclical; the peaks and troughs in vehicle demand essentially parallel economic activity. Even though Chrysler, Ford, and General Motors have diversified through foreign vehicle production and sales, as well as nonautomotive activities in financial services, defense electronics, and vehicle rental companies, vehicle production in North America is an important part of their total sales. The level of vehicle demand is the primary factor in determining profitability, but market share and the mix of cars and trucks and of small and large cars also influence profits.[2]

The domestic automotive industry emerged from the recession of 1980-1982 (when it posted a cumulative loss of $4.1 billion) to a record post-World War II expansion between 1983 and 1989, during which time it earned $61 billion. Restricted Japanese competition between 1981 and 1985, buyer preference for trucks and larger, well-equipped cars, and record vehicle demand in North America combined to create the unprecedented total profit rebound for each of the three domestic automotive companies. (See Figure 5-1.)

Since 1986, the domestic industry has faced increasing competitive pressure. The Japanese automakers have opened eight assembly plants ("transplants") in the United States and thus have been able to add more than 2 million units of locally produced cars and trucks to the 2.3 million imports permitted under the voluntary restraint agreement.[3] Although Japan lost some of its competitive edge when the dollar was depreciated relative to the yen following the 1985 Plaza Accord, the Japanese automobile companies have continued to gain market share in the United States.[4]

---

[2]Labor and overhead costs are relatively constant over all models. Nonetheless, on average, trucks contribute more profit per unit, followed by large luxury automobiles. Trucks are redesigned infrequently, so tooling investment per unit is relatively low. Historically, small cars have been less profitable than large cars, and there has been greater competition in this market segment from imports. Larger cars can command higher prices because of their size and features. Moreover, as discussed herein, the CAFE standards may themselves establish pressures on manufacturers to reduce the prices of small, fuel-efficient cars to ensure compliance.

[3]The Japanese share of the U.S. automotive market rose from 19.6 to 28.1 percent between 1980 and 1990. The Japanese share of the light-truck segment has consistently trailed that for automobiles. In 1980, the Japanese share of light-truck sales was 10.1 percent, rising to 16.1 percent in 1990. The 25 percent tariff on imported two-door trucks limits profitability and, in addition, the Japanese do not, as yet, produce many popular types of light trucks, such as standard-size pickup trucks, vans, and sport-utility vehicles.

[4]In 1985, Britain, France, West Germany, Japan, and the United States met at New York's Plaza Hotel and reached agreement on depreciating the dollar, which made it cheaper for foreigners to buy U.S. goods.

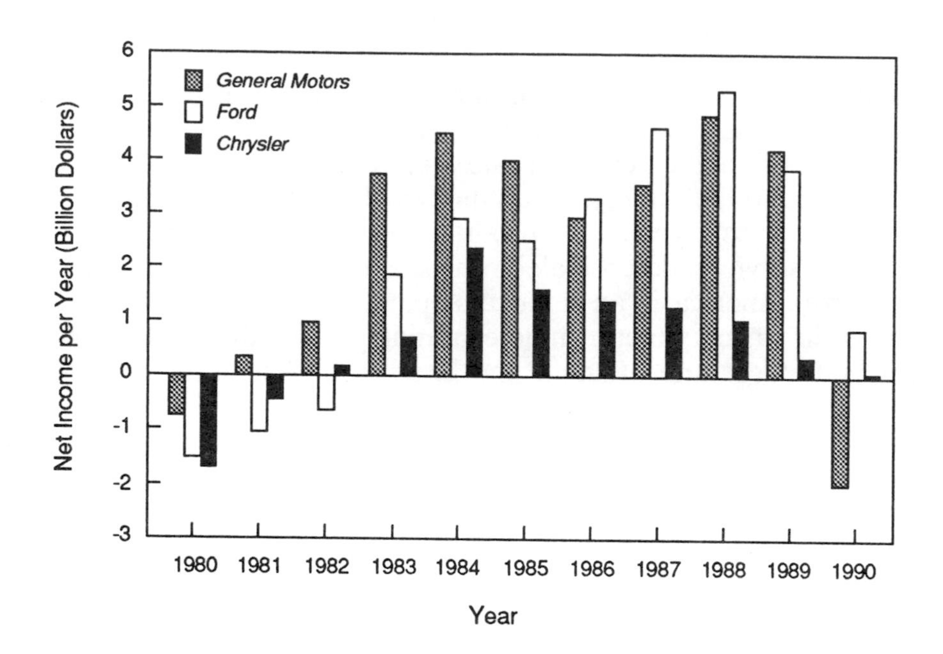

FIGURE 5-1 Net income of Chrysler, Ford, and General Motors automotive companies, 1980-1990.

SOURCE: Based on data from the companies' annual reports.

They retained a tremendous advantage in cost of capital until 1990, as well as an advantage in production efficiency, in factories in Japan and the United States (Womack et al., 1990).[5]

The expansion of Japanese penetration into the automobile and truck sectors of the market has touched off a battle for market share that is a major factor in the recent record losses of the U.S. automakers. On February 24, 1992, General Motors announced corporate losses of $4.45 billion in 1991. General Motor's announcement followed those of Ford and Chrysler reporting losses in 1991 of $2.26 billion and $665 million, respectively. The recession has also had an impact on Japanese automakers, but not as severe as for U.S. automakers. Whereas the U.S. companies have experienced record losses, the principal Japanese automobile companies have experienced only a decline in profits. And the market share of Japanese automobiles has increased to record levels in the United States during the recession.

[5]Their U.S. factories benefit from a young, skilled work force. They have virtually no pension liabilities as yet, and health care costs are low because of the young work force and lack of retirees. They are also not burdened by the need to provide wages and benefits to workers displaced by plant closings. In addition, states that won transplant investment provide incentives in the form of funds for training workers, tax abatements, and so forth.

Capital spending by the domestic automakers was estimated at about $17 billion in 1991, and negative cash flow forced them to raise outside capital, debt, and equity in record amounts in 1991, in addition to cutting dividends. The creditworthiness of all three companies has deteriorated, as evidenced by the credit-rating reductions for all three companies in 1991. As a result, borrowing is now more difficult and more expensive. Although profitability should be restored when an economic rebound occurs, it is doubtful that the domestic industry can match previous levels of income because of such factors as the extensive competition across all model lines, especially those that have been highly profitable to U.S. manufacturers, the extreme overcapacity that exists, and the likely continuation of various incentive programs to purchase or lease cars. Capital spending, on the other hand, will have to remain relatively high to meet government safety and emissions regulations and to respond to Japanese competition, especially Japanese inroads into the markets for larger cars and standard-size trucks.

It is probable that the domestic industry will suffer further loss of market share in the classes of vehicles (large cars and trucks) that contribute a disproportionate share of industry profits even if the overall Japanese share remains stable in the future. The industry's earning power will be further limited in the 1990s by high marketing costs, R&D spending, the interest expenses associated with its record-high debt, and double-digit percentage increases in health care and pension expenses. Moreover, as discussed in more detail in the next chapter, because the population of cars and light trucks is approximately equal to the population of driving age, future demand in the United States will only slightly exceed replacement of scrapped vehicles. With the trend in demand for new vehicles projected at about 16 million units by the year 2000 (cyclical peaks could exceed 16 million, as in 1986), the United States will continue to suffer from severe excess capacity, even during brief periods of strong economic expansion.[6] This is an unprecedented situation.

EFFECTS OF COMPETITION

Industry Trends

The competitive situation of the domestic industry could be adversely affected by any government policy that imposes added burdens on the industry that hit domestic manufacturers harder than foreign manufacturers or that imposes costs significant enough to reduce overall vehicle demand substantially.

Despite worldwide capital spending of $90 billion between 1983 and 1989, the domestic manufacturers lost ground in terms of market share to the Japanese automakers. It is estimated that by 1995 Japan's North American transplant capacity for cars and light trucks could reach 2.8 million units, compared with just over 2 million

---

[6]Based on presentations by Chrysler, Ford, and General Motors to the committee's Impacts Subgroup on September 16, 1991 (see Appendix F).

units in 1991. The combination of Japanese vehicles (imported and locally assembled), imports of European automobiles, and vehicles produced by General Motors, Ford, and Chrysler constitutes a potential supply that could far exceed demand in 1995, even with General Motors' decision to close six more plants. For the foreseeable future, the pressure on profit margins will prompt the withdrawal of some companies and the closing of more domestic capacity.[7]

The North American motor vehicle market remains the most open major market in the world, and thus, it is the target market for foreign manufacturers who wish to expand or shift production. For example, at the same time that some European manufacturers are withdrawing from North America, Korean manufacturers (e.g., Kia) are announcing their intention to sell cars in the United States through an independent dealer network.[8] Europeans see U.S.-made Japanese cars as a means of circumventing their own limits on Japanese cars.

In an attempt to increase sales, the industry has resorted to numerous marketing, incentive, and service programs. These include increasing the period of a car loan to five years (thereby decreasing the monthly payment), offering cash rebates, lowering loan interest rates, subsidizing leases, selling program cars,[9] and offering enhanced warranty and buyer-protection programs. Although these programs are beneficial to the consumer, they are costly to the manufacturers.

Employment Trends

Employment in the domestic automotive industry is likely to continue to decline during the 1990s, regardless of any action on fuel economy regulations. And, industry adjustments to the current overcapacity will further reduce employment.

Domestic automobile manufacturers directly employed 609,800 hourly production workers in 1990. Many more workers are employed in industries that support the

---

[7]In 1991, Sterling (United Kingdom) and Peugeot (France) terminated sales of automobiles in the United States. Other producers might also choose to abandon the market because of the high fixed cost of meeting pending emissions standards. Since 1988, Chrysler has closed 3 assembly plants and opened 1 in the United States, while General Motors has closed 10 assembly plants and opened 1 (see Table 5-1). Another 6 plants and 17 components factories will be closed as a result of capacity cuts announced on December 18, 1991, in addition to 4 other pending assembly plant closings by General Motors. Chrysler will probably close its Toledo plant, which builds the aging Jeep Cherokee. The December 18, 1991, announcement by General Motors was equal to a capacity reduction of 1.4 million units (Frame, 1991).

[8]Kia manufactures the Festiva in Korea, which is sold in the United States by Ford.

[9]Program cars are new cars that are used by automobile rental companies for a short period of time and then repurchased by automobile companies for resale as used cars at a significant loss to the vehicle manufacturer. It is estimated that program cars account for approximately 1.5 to 2.0 million units a year.

TABLE 5-1 Closings and Openings of North American Assembly Plants by the American-Owned Automobile Companies

Company	Plant	Year	Capacity (Car)	Jobs
Closings				
General Motors	Detroit, Michigan	1987	212,000	6,600
General Motors	Flint Body/Pontiac Assembly, Michigan	1987	250,000	4,500
General Motors	Norwood, Ohio	1987	250,000	4,000
General Motors	St. Louis, Missouri	1987	250,000	2,200
Chrysler	Kenosha, Wisconsin	1988	500,000	5,500
General Motors	Leeds, Missouri	1988	250,000	2,200
General Motors	Pontiac, Michigan	1988	100,000	700
General Motors	Framingham, Massachusetts	1989	200,000	2,700
Chrysler	Detroit, Michigan	1990	250,000	1,700 (net)
General Motors	Lakewood, Georgia	1990	200,000	2,200
General Motors	Pontiac, Michigan	1990	54,000	1,500
Chrysler	St. Louis, Missouri	1990	250,000	3,700
General Motors	Flint #1, Michigan	1991	250,000	3,450
General Motors	Willow Run, Michigan	1992	190,000	4,000
General Motors	North Tarrytown, New York	1995	225,000	3,450
Openings				
General Motors (Saturn)	Spring Hill, Tennessee	1990	250,000	3,000
Chrysler	Detroit, Michigan	1992	240,000	3,000 (app.)

SOURCE: Based on information from the automobile manufacturers, Autofacts 1991 Yearbook (Autofacts, Inc. West Chester, Pa.), and the New York Times (February 25, 1992, p. A1).

motor vehicle industry (Motor Vehicle Manufacturers Association [MVMA], 1991; Salter et al., 1985).[10] Hence, any decrease in automotive sales has ripple effects on employment throughout the U.S. economy.

Hourly employment reached a recession low in the first quarter of 1991 as factories closed to respond to low retail sales and inventory reductions by dealers. Apart from the cyclical slump in employment, the industry has lost more than 120,000 hourly jobs since 1978 (see Figure 5-2). Japanese transplants in the United States are blamed for the loss of about 20,000 jobs because they are more efficient and use fewer

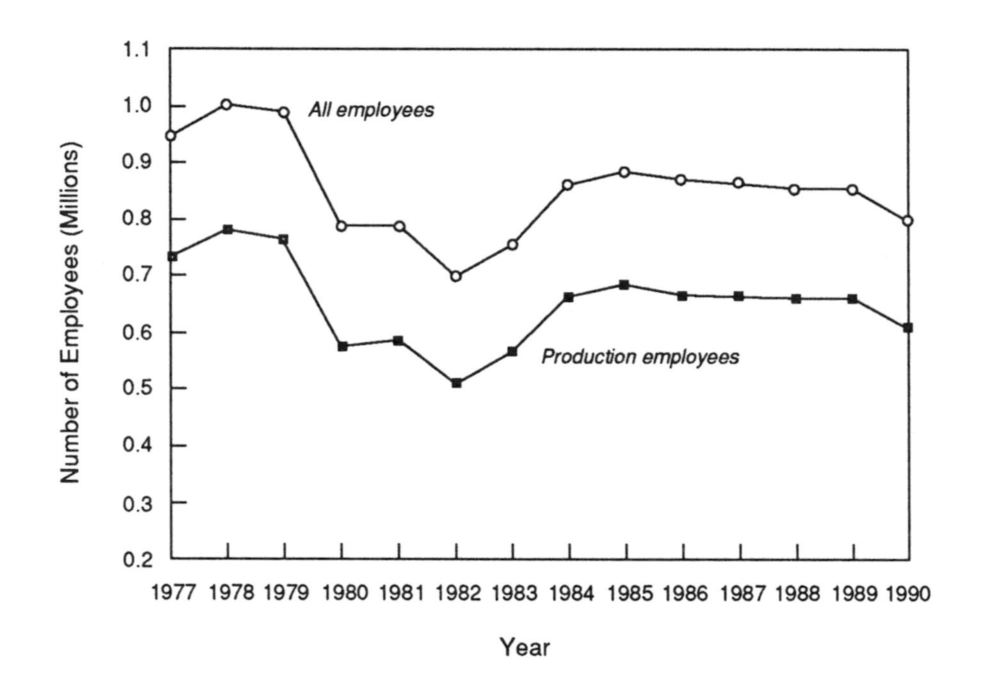

FIGURE 5-2 U.S. motor vehicle and equipment manufacturing employment, 1977-1990.

SOURCE: Based on MVMA (1991).

---

[10]Different analysts report varying employment statistics. Salter et al. (1985) estimated that for every 1 million unit fluctuation in sales, about 150,000 to 200,000 jobs in the automotive and supplier sectors are affected. The U.S. General Accounting Office (GAO, 1988) reported that a typical 1990 U.S. assembly plant had 4,068 assembly workers producing 200,000 vehicles, whereas the United Automobile Workers (UAW) indicated that 5,932 assembly workers would be used in such a plant. Indirect employment is also generated in industries connected to the assembly plants. The GAO (1988) reported that 4.87 jobs were created at auto parts companies and all upstream sectors for every assembly plant worker, whereas it reported a UAW estimate of 4.21 such workers per assembly plant worker. Thus, estimates for vehicles per worker range from 6.5 to about 8.4 units per worker.

locally produced components. Thus, the displacement of one U.S.-made car by a transplant-made car represents a net loss of assembly and components jobs in the United States (GAO, 1988).[11]

In addition to Japanese transplant assembly facilities, approximately 300 to 400 Japanese automotive parts companies have established wholly owned or joint-venture operations in the United States to sell to U.S. and Japanese automotive companies. It is impossible to predict precisely what the impact of these new operations will be. It appears probable, however, that they will rely in part on imported parts and components and will be more productive than their U.S. counterparts, which will result in further job losses. Further inroads by Japanese brands, whether imported or locally assembled, will reduce U.S. hourly jobs. Productivity gains will also reduce labor needs.[12] (Figure 5-3 shows how output per worker has changed over the past 15 years.)

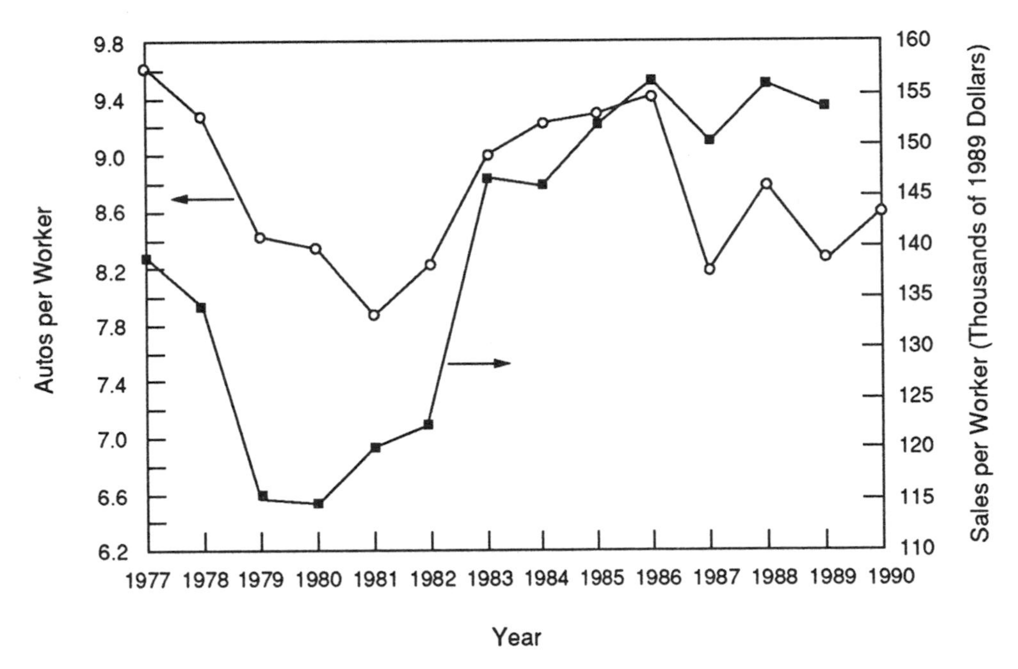

FIGURE 5-3 Net vehicle output (sales of new cars and used cars) per worker in constant 1989 dollars and autos per worker. Employment is that in U.S. motor vehicle and equipment manufacturing.

SOURCE: MVMA (1991).

[11] If the Japanese exported vehicles to the United States rather than producing some vehicles here, the loss of U.S. jobs would have been considerably greater. In its presentation at the committee's meeting on May 13-15, 1991 (see Appendix F), the UAW estimated that for every 200,000 to 250,000 vehicles imported, 30,000 jobs in assembly plants and supplier companies would be lost, or roughly 7 to 8 jobs per car.

[12] The 1990 contract with the UAW allows elimination of one job for every two employees who leave. On December 18, 1991, General Motors said that this would reduce hourly employment by 15,000 people in 1992 and each year through the mid-1990s (Frame, 1991). Attrition, according to General Motors, cut hourly jobs by 25,000 in 1991.

Changing Market Share

As Japanese transplant capacity grew in the United States during the late 1980s, sales and the market shares of Japanese automobile companies increased, to the detriment of domestic companies. In addition to having more cars to sell, Japanese manufacturers followed a logical (particularly considering the import quotas) path of product evolution, capitalizing on the comparable advantage that was available to them (as producers of small cars) under the CAFE system (see Chapter 9). They moved upmarket into larger and more luxurious models by the late 1980s. New divisions (e.g., Lexus, Infiniti, and Acura) were launched to focus on highly profitable luxury cars. The extension of Japanese product ranges into market sectors that provide the bulk of domestic industry profits--mid- and full-size cars and light trucks--has made the Japanese a greater threat to domestic industry profitability in the future than it has been over the past 10 years. For example, although the Honda Accord and Toyota Camry were initially introduced as compact cars, both have increased in size and luxury features and are currently classified as midsize cars. The substitution of a $30,000 Japanese car for a similarly priced domestic model has severe profit implications for domestic automakers.

The Japanese producers have also fragmented the U.S. market by expanding their product offerings into low-volume specialty and luxury models. Today, the American consumer can choose among approximately 290 nameplates of cars and light trucks, compared with 170 only 10 years ago.[13] The domestic automotive industry has traditionally relied on large volumes of a single model to attain production efficiency. The Japanese, on the other hand, have very efficient, flexible plants capable of producing several models and of adjusting to changes in the marketplace (Womack et al., 1990). As a result, the Japanese automakers have evolved from an initial strategy when they first entered the U.S. market of concentrating on "econobox" utilitarian vehicles to the current strategy of extensive product diversity. This strategy has enhanced the threat they pose to the domestic manufacturers.

Product Development

The Japanese automakers also have a more efficient product-development process than their U.S. counterparts although there is evidence that this situation is changing. Whereas it takes U.S. automakers about 60 months, on average, to develop a new product, Japanese automakers take only 47 months. Moreover, the product-development effort requires 3 million person-hours in the United States, in comparison with only 1.7 million person-hours in Japan (Clark and Fujimoto, 1991). Japanese automakers thus have a significant competitive edge in product development with respect to development time and resource requirements.

---

[13]Between 1982 and 1990, Japanese automobile companies doubled the number of model offerings from 45 to 90, whereas the United States had a modest increase--from 35 to 45 models. In 1976, the top-selling car model in the United States had sales of 621,140, whereas in 1989 the top-selling model, the Honda Accord, had sales of only 362,707. The top 10 models had total sales of 2.78 million cars (28 percent of the new-car market) in 1989 versus 3.58 million (32 percent of the new-car market) in 1978 (Wards Communications, various years).

The Japanese automobile industry has followed a product-replacement philosophy based on short product cycles (roughly half the time of the U.S. companies) and fast depreciation of assets. Short product cycles mean that vehicles can be adjusted to changes in the marketplace more frequently. Features that are more easily incorporated with major changes can be introduced more rapidly. Tools and equipment that are model specific can be written off over the short life of the model.[14] Because domestic manufacturers have followed longer product cycles, they are less nimble in the marketplace than their Japanese competitors.[15]

The best Japanese automakers have developed a production system that enables them to produce efficiently at a smaller scale than the American manufacturers while manufacturing products of high quality (Womack et al., 1990). The annual production volume for the Japanese product cycle can be economically viable at levels as low as 50,000 units, compared with optimum production levels of 200,000 units for a U.S. model. The four-to-one difference in production volume enables Japanese automakers to provide four times as many product offerings from a single plant as a U.S. manufacturer, thereby providing the opportunity to satisfy more market segments.

The relationship between Japanese companies and their principal suppliers may also provide a competitive advantage for the largest companies in product development involving new technology. Japanese automobile manufacturers enter into long-term relationships with their principal suppliers and are bound together in business groups through joint-equity relationships termed *Keiretsu*. As a result of these long-term relationships, joint product development is possible. In the United States, suppliers are involved jointly in only 14 percent of the engineering effort in new product development, whereas in Japan suppliers account for 51 percent of that engineering.[16] U.S. automotive companies, especially General Motors and to a lesser extent Ford and Chrysler, are vertically integrated more extensively than the Japanese manufacturers,

---

[14]Japanese firms also write off investments on key components, like engines and transmissions, more quickly than American firms. Toyota, for example, launched 14 new engines between 1978 and 1990, fully depreciating its engine facilities over six years. Engine and drivetrain life is at least 12 to 15 years for domestic companies; Ford, for example, did not launch a single new engine during the 1980s.

[15]For example, the Honda Accord was redesigned in 1982, 1986, and 1990. In comparison, the Ford Taurus was introduced in 1985, was significantly updated in 1991, and is scheduled to be face-lifted in 1995 before being completely changed in 1999 (based on presentations to the committee by Ford Motor Company). Thus, the Accord had two major redesigns in an 8-year period (1982-1990), whereas the Taurus is scheduled to have two major redesigns over a 14-year period. Chrysler's first completely new automobile platform since 1982 will be launched in the fall of 1992.

The Japanese product-development strategy of small-scale production and extensive product diversity also has implications for the introduction of new technology. It allows Japanese companies to experiment with new technology on small-volume products before committing more broadly to the technology on many product lines. Moreover, because it is easier and more cost-effective to incorporate new technologies at the time of a facelift or model change, the Japanese will retain the ability to adopt some technologies faster than their American competitors.

[16]The committee distinguished subcontracted engineering effort from joint R&D. The latter may provide a competitive advantage because it allows the manufacturer to augment its own technical skills with those of its supplier.

which does provide economies somewhat analogous to the Japanese *Keiretsu* system. Nonetheless, even considering this fact, the Japanese have been more effective at joint product development, particularly in cooperation among different industries as well as within *Keiretsu* structures.

Manufacturing Productivity

Although the American automobile industry has made some progress in introducing flexible manufacturing, it has failed to match Japanese levels of manufacturing and organizational efficiency. Improved factory operating efficiency and vehicle design have yielded higher quality and productivity, but that only partly closed the competitive gap. A recent study (Womack et al., 1990) comparing the performance of auto assembly plants throughout the world found that U.S. plants operated by the U.S. automakers had an average productivity of 24.9 hours per vehicle. In comparison, Japanese plants in Japan had an average productivity of 16.8 hours per vehicle. Cars assembled at the U.S. plants had 7.8 assembly defects per car in comparison with 5.2 defects per car in the Japanese plants. The gulf between American and Japanese manufacturers is narrowing but remains a troubling national problem.

Capacity for Investment

Access to cheap capital through the issuance of warrants and convertible bonds worth billions of yen during the late 1980s supported Japanese investment in new plants and products.[17] While the Japanese automobile companies must replace this low-cost debt with more expensive capital, they continue to benefit from the investments made during the late 1980s.

Although the domestic automakers also invested heavily during the 1980s, much of the surplus cash was used for acquisitions outside the North American automotive market. In addition, cash was also used to repurchase shares in attempts to boost stock prices, on products and facilities for overseas subsidiaries, or on domestic facilities deemed necessary for the future of the business, such as the billion-dollar Chrysler Technology Center. Despite record spending on domestic automotive operations, the industry must still invest to match Japanese production standards and to respond to the shift upmarket (i.e., the movement to larger, more expensive cars) of Japanese manufacturers. The prospect of lower vehicle demand, further losses in market share, and significantly higher fixed costs raises questions as to the domestic industry's capacity to fund investments to close the competitive gap with the Japanese and to meet the safety and emissions standards that have been enacted. Investments in fuel economy technologies that force the early retirement of models or components, or that must be accomplished outside the planned spending cycles, could place an untenable financial burden on domestic automakers.

---

[17]Convertible debentures and warrant bonds were sold by Japanese companies to public investors, often at interest rates of 1 to 2 percent. This enabled long-term projects to be financed despite low returns on investment. In contrast, the long-term U.S. interest rates to automotive borrowers at that time were about 10 percent.

The automobile industry is characterized by long lead times for major component systems, such as engines and transmissions, and complete vehicles. To determine the design and features of a vehicle or of critical components, a two- to four-year lead time is necessary for ordering specialized equipment and preparing manufacturing plants. Vehicle assembly facilities require a lead time of two to four years. (The required lead time is affected by the extent of change in the vehicle and the production process.) Three to four years are required to prepare for an all new engine or drivetrain (U.S. Department of Transportation, 1991).

Some American automotive companies seemed slow to convert to new models and contemporary engine designs during the 1980s. However, every automotive company is expected to spend a relatively large amount of capital on new products and facilities well into the 1990s. In the process, a significant number of new models and four-valve-per-cylinder aluminum engines will be launched. Because these new models and components are already set, the industry's ability to adapt to more stringent fuel economy standards prior to MY 1996 is limited. Nonetheless, there will probably be some improvement in fuel economy in these newer vehicles over their current counterparts.

The industry relies largely on internally generated funds to provide the bulk of capital for investment. As a result, capital spending should roughly match depreciation and amortization of equipment and tools. The long life typical of domestic car lines (10 years or more) and of major components (10 to 15 years) results in a slow write-off of assets. The product-development and depreciation strategies of domestic automakers--as well as significant spending into the early 1990s--are a potential obstacle to investment for the rapid introduction of new technologies beyond those already planned. If fuel economy standards are enacted that make vehicles, production equipment, or key components obsolete before their normal retirement, the industry's financial condition will be compromised.

Nonetheless, a long-term fuel economy standard or goal should be within the industry's capacity, as long as it does not distort normal product evolution. Within 10 to 15 years, all current models and most engines and drivetrains will undergo at least one major change, and the equipment used in their manufacture will be written off. If fuel economy standards follow the industry's own product-development process, some, but not all, of the financial risk of the new standards would be reduced.

It is difficult to comment about the ability of the automotive parts and materials industry to invest in fuel economy technologies. Once again, over a long period of time, such investments could be incorporated into the normal product programs of these companies. If such investment created new opportunities--for example, in the application of aluminum and plastics--the beneficiaries would support necessary development and engineering efforts. Government regulation has hurt some parts makers in the past, but it has also created new industries--for example, producers of airbags, catalytic converters, and fuel-injection systems. No doubt fuel-saving technology would make some materials, processes, and existing systems obsolete, but their replacement by other materials, components, and technologies could create new opportunities for other companies. Again, however, the adverse financial impact would be minimized by allowing the transition to occur at a natural pace.

Structural Change in the Industry

The U.S. automotive industry is in the midst of a major transition. Recent studies have documented that the best Japanese companies not only perform better than their American counterparts, but also operate in a fundamentally different manner. On average, the best Japanese companies can manufacture cars of better quality and with superior manufacturing productivity compared with American manufacturers but the gap is narrowing.

The best Japanese automobile companies have redefined the bases of international competition and introduced new standards of "best practice." They have done this by introducing an approach to manufacturing and product development that is fundamentally different from the conventional mass-production system that has been the basis of U.S. automotive manufacturing. This new manufacturing and product-development approach, commonly referred to as lean production, is based on a different concept of work organization and human resource utilization within and between organizations. If the U.S. automotive industry is to remain competitive, it must undergo structural change and adopt this new production paradigm. The process has already begun, but it is time consuming, complex, expensive, and demanding. The industry has limited capacity to accommodate other changes during this period of intensive competition and transition.

In addition to the structural changes, the American automotive industry must confront a series of regulatory requirements for safety and emissions control in the 1990s. There is also the prospect of international regulation of greenhouse gases. Care should be taken in placing additional burdens on an industry with a full agenda.

No doubt some people will claim that many of the U.S. industry's problems are of its own making. The industry might be said to have waited too long to respond to competitive pressures, to have made an inadequate response, and to have chosen inappropriate priorities in its allocation of human and financial resources. Whether or not that is the case, the industry is a significant sector of the U.S. economy and it is in serious trouble. Fuel economy standards above those demanded by the marketplace should be evaluated with this reality in mind.

THE INTERNATIONAL AUTOMOTIVE MARKET

Automotive manufacturing is a global industry and most companies seek to sell their products outside their home markets. U.S. companies produce motor vehicles through subsidiaries in Europe, Australia, Latin America, and elsewhere, that are different from the vehicles they produce in the United States. American automobile companies are thus directly affected by the energy and emissions policies of other major automobile-producing countries.

Impact of Fuel Prices

Traditional American-built cars have been disadvantaged in major foreign markets, such as Japan and Europe, that have used high gasoline taxes to accomplish energy-conservation goals and to raise tax revenues. Because they have evolved under a regime of low fuel prices, American cars have been too large and inefficient to compete, even if the official barriers to their sale in some countries were eliminated. As a result, the European subsidiaries of Ford, General Motors, and (formerly) Chrysler, developed families of automobiles that conformed to the requirements of their host countries. Even though American cars have improved significantly in terms of fuel economy over the past 15 years, they still carry the reputation in foreign markets of being gas guzzlers. In effect, cheap energy has been one factor that has confined American cars to the North American market.

In contrast, Japanese and European manufacturers have exported their vehicles, developed in markets with high fuel prices, to markets throughout the world. Their highly fuel-efficient cars were salable in all countries, whether fuel was expensive or cheap. For example, foreign brands account for about 35 percent of the U.S. market. Whereas American manufacturers have developed separate product lines for domestic and foreign customers, the Japanese and Europeans have been able to sell the same models worldwide, which gives them a significant economic advantage.

Because gasoline prices are higher overseas, it could be easier for Japanese and European companies to introduce costly fuel-saving technologies more quickly throughout the world than for American companies. For example, Toyota has established a policy of reimbursing suppliers 1,000 yen per part or component for every kilogram of weight saved (about $3 a pound). This "bounty" is cost-effective for the Japanese consumer because, at the cost of fuel in Japan, a consumer can recover the resulting increase in the vehicle price through lower expenditures on fuel (U.S. Department of Commerce and Motor Equipment Manufacturers Association, 1990).

Canada has had higher gasoline prices than the United States for more than a decade and currently has gasoline prices about 60 percent higher than in the United States. Canadian fuel economy standards rose from about 19.7 mpg in 1978 to 29 mpg in 1990, compared with 19.9 to 27.8 mpg in the United States over the same period. On August 1, 1991, Canada also enacted the Tax for Fuel Conservation to encourage the purchase of fuel-efficient cars.[18]

---

[18]Consumers will be paid $100 (Canadian) for vehicles purchased with fuel economy better than 6.0 liters per 100 kilometers (about 39 mpg). The legislation imposes a tax ranging from $75 for cars using 6.0 to 9.0 liters per 100 kilometers (about 39 to 26 mpg) to $7,000 on cars consuming 18.1 liters per 100 kilometers (13 mpg). Sport-utility vehicles are also covered under this law; the taxes range from $75 for vehicles using 8.0 to 9.0 liters per 100 kilometers (about 29.4 to 26 mpg) to $3,200 for vehicles using more than 18.1 liters of fuel per 100 kilometers. The Canadian approach to energy conservation will result in a mix shift through direct encouragement of the public to purchase higher fuel economy vehicles.

Concerns Over Greenhouse Gases

Europe appears to be ahead of the United States in addressing emissions of greenhouse gases. The Commission of the European Community is drafting legislation, after more than two years of debate, proposing that an energy tax equivalent to $10 a barrel of oil be imposed by the year 2000 (Clean Coal/Synfuels Letter, 1992). The tax would start at $3 a barrel in 1993. At this time it is not known whether other taxes on fuel would be reduced, but it is clear that the intent of the European Commission is to send a message to industry and consumers that "environment costs must be internalized" (Wolf, 1991).

European automobile companies have voluntarily agreed through their Association des Constructeurs Europeens d'Automobiles (ACEA) that they will reduce the carbon dioxide emissions of their vehicles by 10 percent between 1993 and 2005. This means that manufacturers intend to achieve a 10 percent improvement in fleet fuel economy. The ACEA has indicated that it would achieve the targeted fuel economy improvement over that period by forcing "changes in vehicle selection patterns by influencing the consumer's decisions when purchasing a car." In addition, the ACEA will cooperate with other programs to limit carbon dioxide emissions of the entire fleet through reduction in traffic congestion.

In a separate action, the German automobile industry has agreed to a 25 percent increase in the fuel economy of the fleet of new cars on the road between 1998 and 2005 if the government institutes traffic management policies that will add another 10 percent benefit.[19] The German industry has made the assumption that taxation will increase the price of fuel to three deutsche marks per liter (about $7 a gallon) and, with that incentive to the consumer for favoring gas-saving technology, consumers might be more willing to accept a variety of technologies, including direct-injection diesels and engine on-off, among others, to accomplish the goal.[20]

FINDINGS AND CONCLUSIONS

- The domestic automotive industry is in a period of unprecedented financial challenge. The current recession, the intense competition with the Japanese manufacturers, and soaring costs mean that the domestic manufacturers confront serious financial burdens, wholly apart from any burdens associated with changes in fuel economy regulations.

---

[19]Note that the normal turnover of the U.S. fleet, if new car automotive fuel economy remained in the range of 26 to 27 mpg, would also result in significant increases in the fuel economy of the existing fleet.

[20]Engine on-off refers to an engine that shuts off rather than idles. When the driver depresses the accelerator, the engine starts up again. Volkswagen is concerned about adverse consumer response to this technology. Eventual acceptance may occur, however, as drivers become confident that the technology is reliable, that it does not interfere with normal vehicle performance, and that there is an environmental benefit as well as an economic one. The high tax on gasoline no doubt serves an important role in prompting acceptance of such new technologies.

- Employment in the U.S. automotive industry has declined significantly and the trend is likely to continue during the 1990s. The world automotive industry, particularly the domestic industry, suffers from overcapacity, and further plant closings and reductions in employment are inevitable. These changes will have ripple effects throughout the U.S. economy.

- The market share of Japanese automotive manufacturers has grown and is expected to continue to grow, especially in the size classes (trucks and large cars) that are important to the profitability of the domestic industry. Unlike the domestic manufacturers, the Japanese have very efficient, flexible plants capable of producing several models and of adjusting rapidly to changes. As a result, the Japanese manufacturers are more nimble in the marketplace than the domestic manufacturers.

- The product development and investment cycle of the automotive industry requires long lead times and huge amounts of capital. The manufacturers are in the midst of a major wave of investment in new models and major engine plants, which will make it difficult for them to modify the fuel economy that is planned for vehicles to be launched through MY 1995.

- The domestic automotive industry relies on internally generated funds to provide the bulk of capital for investment in new plant and equipment. Within 10 to 15 years, all current models and most engines and drivetrains will undergo at least one major change, and the equipment used in their manufacture will be written off. If the timing of new fuel economy standards follows the industry's product-development schedule, some, but not all, of the financial risk of new standards would be reduced.

- The domestic automotive industry is in the midst of major transition. If the industry is to remain competitive with the Japanese manufacturers, it must undergo structural change and accommodate itself to new production methods. That aside, the industry has limited capacity to accommodate other changes during a period of intense competition. The imposition of new fuel economy standards should be evaluated with this reality in mind.

- The automobile industry is global in nature, and domestic manufacturers will be directly affected by overseas initiatives that have an impact on fuel economy. Fuel prices are much higher in Europe and Japan than in the United States, which creates incentives for the automobile industry there to pursue fuel economy improvements aggressively.

REFERENCES

Clark, K.B., and T. Fujimoto. 1991. *Product Development Performance*. Cambridge, Mass.: Harvard Business School Press.

Clean Coal/Synfuels Letter. 1992. EC eyes carbon tax for possible enactment by 1993. January 13:5.

Frame, P. 1991. Stempel's GM: 80,000 fewer jobs. *Automotive News* December 23:1.

Motor Vehicle Manufacturers Association (MVMA). 1991. *Facts and Figures '91*. Washington, D.C.

Salter, M.S., A.M. Webber, and D. Dyer. 1985. U.S. competitiveness in global industries: Lessons from the auto industry. In *U.S. Competitiveness in the World Economy*, B.R. Scott and G.C. Lodge, eds. Cambridge, Mass.: Harvard Business School Press.

U.S. Department of Commerce and Motor Equipment Manufacturers Association. 1990. Weight reduction offensive spurs industry activity. *JAPAN, News and Analysis for the U.S. Vehicle Parts Industry*. August/September:1.

U.S. Department of Energy. 1992. *Gas Mileage Guide*. Document DOE/CE-0019/11. Washington, D.C.

U.S. Department of Transportation. 1991. *Briefing Book on the United States Motor Vehicle Industry and Market*, Version 1. Cambridge, Mass.: John A. Volpe National Transportation Systems Center.

U.S. General Accounting Office (GAO). 1988. *Foreign Investment: Growing Japanese Presence in the U.S. Auto Industry*. GAO/NSIAD-88-11. Washington, D.C.

Wards Communication, Inc. Various years. *Wards Automotive Yearbook*. Detroit, Mich.

Wolf, J. 1991. Support grows within the EC for energy tax. *Wall Street Journal* December 16:B6C.

Womack, J., D. Jones, and D. Roos. 1990. *The Machine That Changed the World*. New York: Rawson Associates.

6

THE AUTOMOBILE, FUEL ECONOMY, AND THE CONSUMER

To understand how the automobile industry might be affected by fuel economy standards, it is important to understand the automobile consumer. No matter what the automobile companies do, the consumer must voluntarily agree to purchase from the menu of choices offered. This chapter examines what motivates the consumer, how consumer choices have changed over the past two decades, and how one might evaluate the impact of fuel economy standards on the consumer.

THE AUTO MARKET IS A REPLACEMENT MARKET

As mentioned in Chapter 5, the U.S. automotive market is mature; to a very large extent, consumers purchase new cars to replace cars that have worn out. Figure 6-1 shows the additions and subtractions (sales and scrappage) of cars and light trucks in the United States between 1970 and 1989. The scrappage numbers represent automobiles, trucks, and buses, with automobiles accounting for about 80 percent in 1989. Throughout the period, the ratio of licensed drivers to automobile registrations remained virtually unchanged at a value of 1.2; the ratio of licensed drivers to automobiles on the road has changed from 1.37 in 1978 to 1.34 in 1989 (Motor Vehicle Manufacturers Association [MVMA], 1991). Figure 6-2 shows registrations and automobiles on the road in recent years, as well as the U.S. population between 16 and 65 years of age, historically and projected through 2020. The size of the latter group--approximately the set of licensed drivers--is projected to grow from 1991 to 2010, level off at 2010, and then decline.

While sales growth in terms of number of vehicles has largely leveled off, sales in terms of value have not. Figure 6-3 shows the price of new cars and median family earnings, expressed in constant 1970 dollars, for the period 1971-1989.[1] Three prices are shown. One is the average new-car price--it clearly went up over the period, but

---

[1]In nominal terms, the average transaction price on a new car rose from $4,950 to $16,012 between 1975 and 1990.

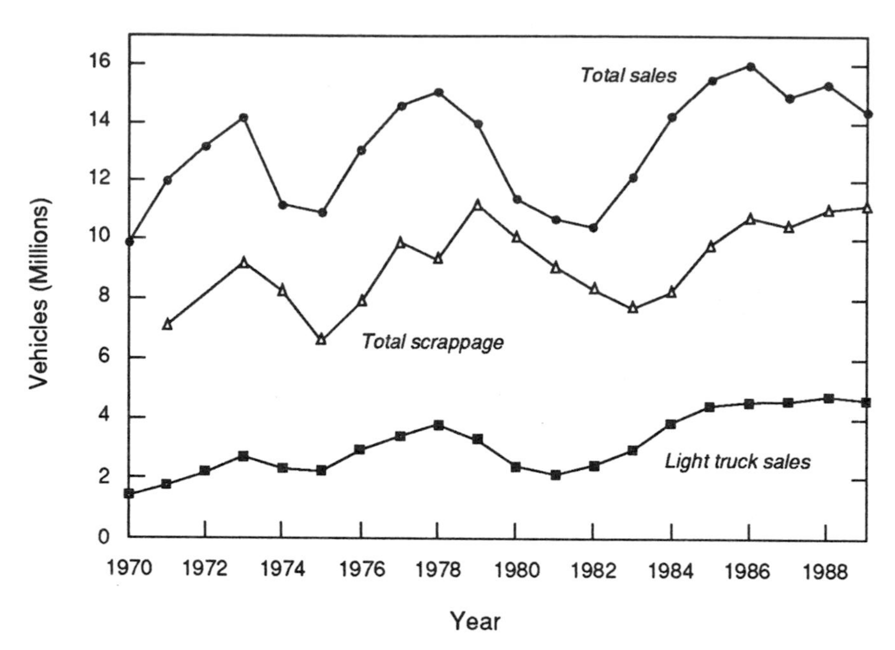

FIGURE 6-1 Trends in the sale and scrappage of automobiles and light trucks.

SOURCE: MVMA (1991).

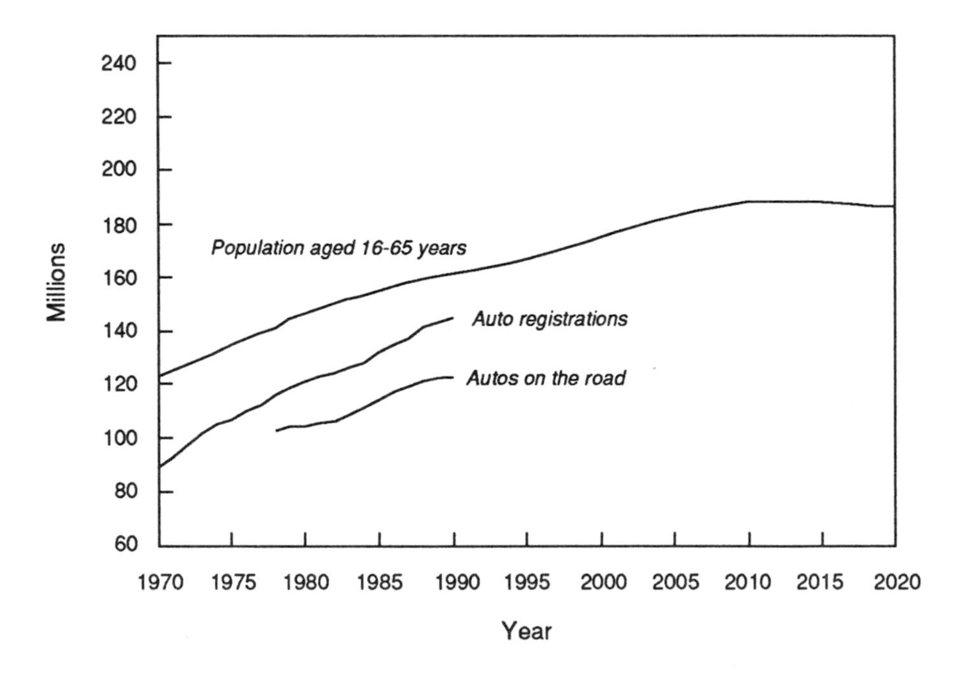

FIGURE 6-2 U.S. population between 16 and 65 years of age, historical and projected, and automobile registrations.

SOURCE: Bureau of the Census (1989, 1991).

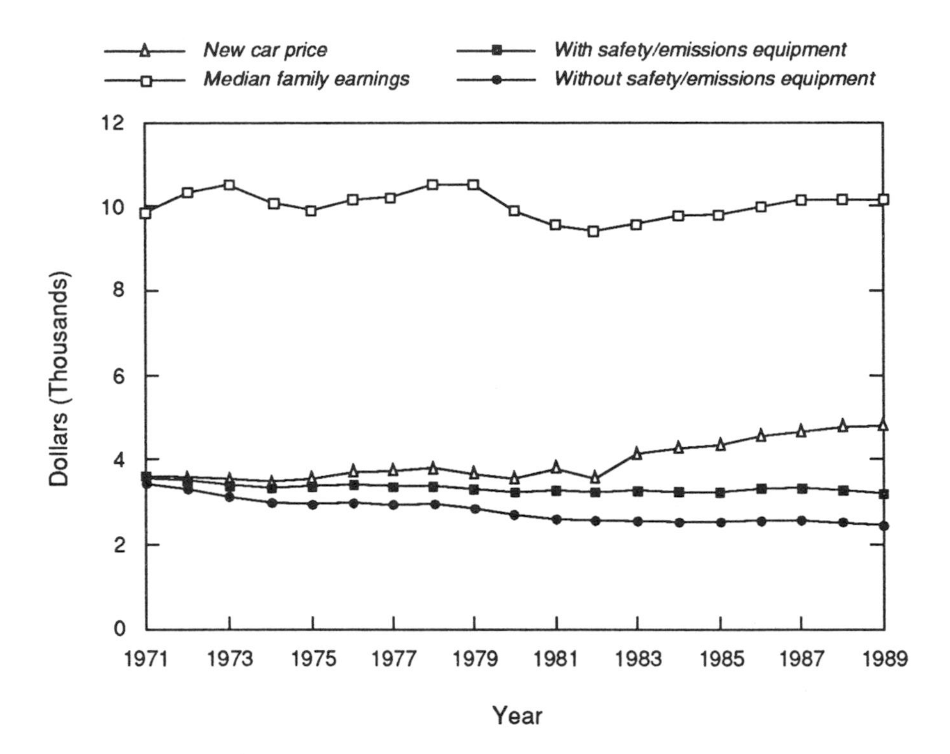

FIGURE 6-3 Family income and new-car expenditures in constant 1970 dollars. The line labeled "without safety/emission equipment" represents a 1967-comparable car. The line labeled "with safety/emissions equipment" reflects the impact of such equipment on the cost of a 1967-comparable car.

SOURCE: Based on MVMA (1991), corrected for inflation using the Consumer Price Index.

incomes did not. Also shown is the price of a "constant-equipped" car, often called a 1967-comparable car. (Consumers cannot typically buy a 1967 constant-equipped new car, but the figure provides an estimate of what it would cost.) The third price shown is the price for a constant-equipped car, but with the cost of the emissions and safety equipment that has been required over time added on. The price of 1967-comparable cars has dropped significantly over the past two decades, even including mandated emissions and safety equipment. But consumers have chosen to spend more and more on their vehicles.[2] Although some of the increase in expenditure may have been to attain additional fuel economy, undoubtedly much of it was to purchase greater performance, quality, and comfort.

---

[2]Some of the increase in expenditure may be due to the movement in the industry away from pricing options individually and to selling options as packages.

Increased expenditures on the automobile "package" should not be viewed as frivolous enhancement of otherwise totally adequate vehicles. Consumers view cars in a substantially different way today than they did several decades ago. The car is more than just a means to get from one place to another. The addition of comfort options to a car makes it a much more functional, multipurpose environment, which is particularly appropriate as people spend more and more commuting time in vehicles. Further, consumers have demanded some features that are societally desirable, such as improved occupant safety. Quality has also risen, which justifies additional expenditures. Other features have been mandated by government to improve air quality, safety, and fuel economy. And others are of a performance nature. Over time, demand is expected to continue to grow for some of these options (such as safety). The demands, however, may be in direct conflict with those for higher fuel economy.

New car buyers today are typically replacing a relatively new vehicle, usually one that is three to six years old, which they trade in toward the new car. It is easy for them to defer buying new vehicles, which is one reason for the cyclic nature of automobile demand. The car traded in becomes transportation for another driver; each vehicle has about three owners before it is scrapped. More old vehicles are on the road than has been the case in the past. The portion of cars and trucks in the United States over the age of 12 climbed from 11.9 to 20.8 percent for cars and from 18.4 to 27.6 percent for trucks between 1980 and 1990 (MVMA, 1991). Whatever the explanation for the increasing age of the fleet, the result is that additions of new cars decrease as the average age of the fleet increases; slow scrappage of cars leads to fewer purchases of new cars.

There is some suggestion that consumers will not continue to increase their demand for optional equipment and performance. Many consumers are faced with imperfect capital markets in the sense that the size of monthly payments on a new car loan (amortized over less than the car's lifetime) is a constraint on how much they can spend on a new car.[3] Over half of all new-car purchases are financed with new-car loans (62 percent in 1990; MVMA, 1991). Figure 6-4 shows the average monthly payment for those who financed their new-car purchase between 1979 and 1989. Because of an increase in the maximum possible length of the term of the loan (from four to five years) in the early 1980s, the average monthly payment has not gone up as rapidly as new-car prices. The maximum loan period is market driven, that is, it is associated with the rate of depreciation of the automobile, which in turn is influenced in part by its lifetime and durability and the demand for the automobile in the used car market. For obvious reasons, the outstanding principal on an automobile loan should not, in general, exceed the value of that automobile in the used car market. In some cases, five-year loans have allowed borrowers to go "upside-down" in their loans--the outstanding balance on the loan exceeds the value of the automobile--for much of the loan term. This is obviously not desirable from the lender's perspective

---

[3]That the size of monthly payments is a "constraint" on the ability of some consumers to buy a new car was expressed to the committee by several of the automobile manufacturers.

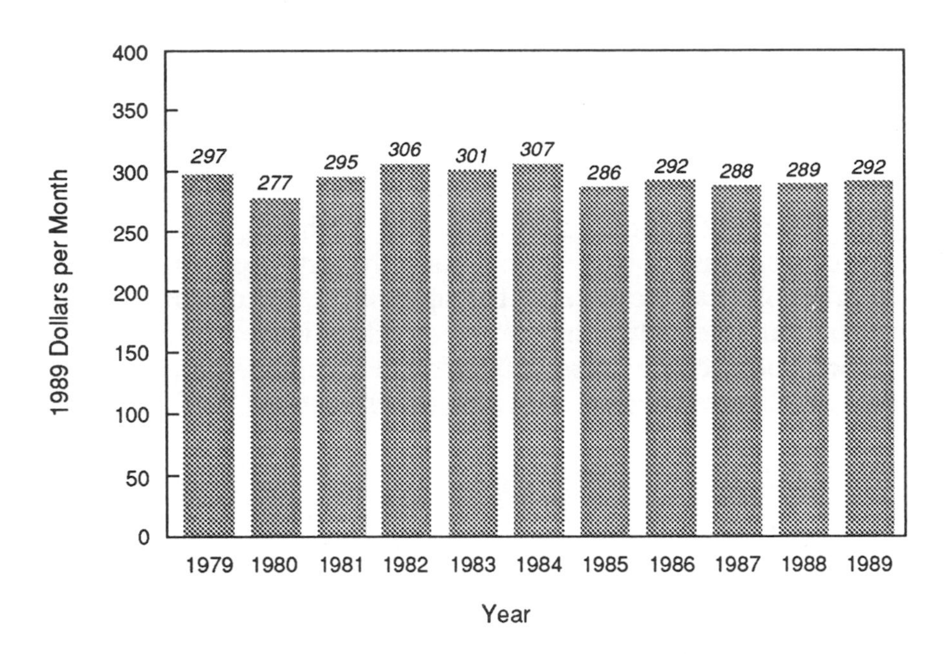

FIGURE 6-4 Average monthly payment for new-car loans (1989 dollars).

SOURCE: Based on MVMA (1991).

and has limited the lenders' willingness to extend loan terms further. As a result, further price increases in excess of the growth in income will be more difficult for consumers to handle than were the increases in the 1980s.

In sum, analysis of the available information suggests that the total number of sales of new vehicles will not grow significantly between now and 2006. Moreover, although the prices of new cars have increased significantly in constant dollar terms since the 1970s, there are reasons to believe that future price increases will not be readily accommodated by consumers.

BALANCING FUEL COST, PURCHASE PRICE, AND VEHICLE CHARACTERISTICS

In purchasing a motor vehicle, consumers consider a wide variety of attributes, including fuel economy, performance, safety, interior volume, and accessories. Fuel economy provides virtually no direct benefit other than decreasing the cost of driving.[4] When the price of fuel increases, consumers are willing to pay more for fuel economy; when the price drops, consumers are less willing to pay for fuel economy. The price of gasoline thus influences the automobile purchase decision. Consumers evaluate how much fuel economy to buy in a new car and how much of other vehicle characteristics to buy (e.g., safety, performance).

Figure 6-5 illustrates how the rational consumer trades off the increased cost of technology to improve automobile fuel economy with the resulting fuel savings. Holding all other characteristics of a vehicle constant, line 3 on Figure 6-5 shows how the hypothetical price of a car increases with improved fuel economy (miles per gallon, or mpg).[5] (The qualitative conclusions drawn from the figure are not very sensitive to specific assumptions about cost.) Also shown in the figure is the present value of the lifetime stream of gasoline expenditures at two gasoline prices, $1.00 a gallon and $3.00

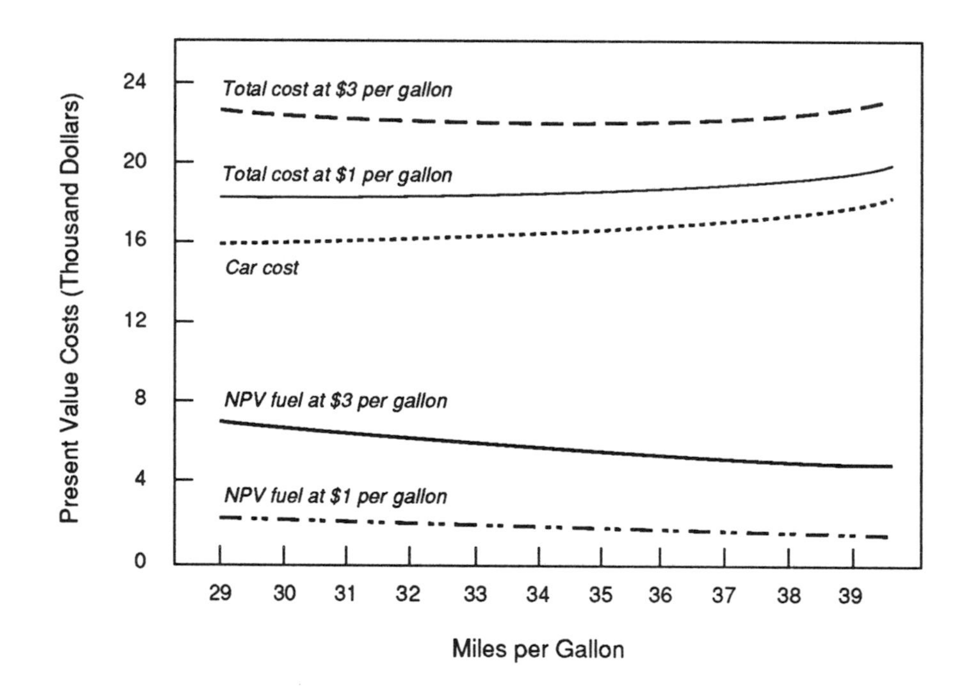

FIGURE 6-5 Hypothetical total car and net present value (NPV) fuel costs (assumes 10,000 miles traveled per year, a 10-year automobile life, and a discount rate of 10 percent a year).

---

[4]Consumer interest in fuel economy is also affected by the availability of fuel. The gasoline lines of the 1970s created a wave of consumer interest in small, high fuel economy cars. Interruption in gasoline supply thus can have an immediate impact on buyer interest in fuel economy.

[5]The curve starts at the 1990 average price of a new car ($16,012) (MVMA, 1991) at a fuel economy of 29 mpg. Fuel economy is assumed to increase to 39.5 mpg at an additional cost of $2,500.

a gallon (lines 1 and 2).[6] The purchase price and gasoline costs are combined into a curve of total costs (excluding other operating costs) in lines 4 and 5 (corresponding to the two gasoline prices).

Figure 6-5 illustrates three points. First, the fuel economy level that results in the lowest total cost (the optimum for the consumer) is not too dissimilar for the two gasoline prices (30 mpg for $1.00 per gallon and 33 mpg for $3.00 per gallon). This suggests that gasoline price is not a very powerful lever for increasing the efficiency of the new-car fleet.

Second, the total cost curve is very flat in the vicinity of the optimal fuel economy level--total costs are relatively insensitive to the fuel economy level of the automobile. Thus, the consumer would not be expected to devote much effort to determining the best level of fuel economy to acquire. For curve 5, the difference in total cost between a 33-mpg car and a 39-mpg car is approximately $1,000, less than 5 percent of total costs. Because the economic benefits of improved fuel economy are not dramatic, consumers are behaving rationally in not demanding improved fuel economy, and manufacturers understandably may be reluctant to take significant risks to develop more fuel-efficient vehicles. Indeed, as will be discussed in Chapter 9, the technology-forcing dimension of CAFE-style regulation is one of its principal benefits.

Finally, because of the nature of the fuel economy measure (in miles per gallon), the total fuel bill does not drop as dramatically as one might expect as fuel economy moves from 29 to 40 mpg. This is because fuel use is proportional to the reciprocal of miles per gallon. Thus, the gasoline saved in going from 20 to 25 mpg is about the same as that saved in going from 30 to 43 mpg.

Gasoline price not only affects the consumer's interest in fuel economy, but also his or her interest in other automobile characteristics. When the fuel price dropped substantially in the mid-1980s, the cost of owning and operating a car dropped, as well as the additional cost associated with increased performance (since performance is inversely related to fuel economy). In essence, there are many features a buyer would like to include in his or her vehicle--safety, roominess, maneuverability, creature comforts, acceleration, speed, and durability. All these features add to the initial price or operating costs (particularly fuel) of a car. Inevitably the consumer makes trade-offs. As fuel prices drop, the consumer values fuel economy less and may buy more of these other features.[7]

---

[6]The figure assumes 10,000 miles per year of driving, a 10-year automobile life, and a discount rate of 10 percent per year.

[7]Another factor affecting consumer purchases is disposable income. As incomes rise, people may wish to buy more automobile (not necessarily more units, but more in terms of features). If an automobile is a luxury good in the sense that as incomes rise, people spend proportionally more on automobiles, then demand can be expected to rise even more rapidly than incomes. In any event, if incomes go up in the United States, one can expect to see increased demand for automobiles, in terms of value.

Gasoline price also affects vehicle usage. When the price of gasoline increases, the cost per mile of transportation services rises. The consumer can and will respond by reducing miles traveled. Figure 6-6 shows annual vehicle miles traveled per vehicle versus the price of gasoline. Although this figure does not reflect other factors that influence miles traveled, it does show the effect on miles traveled of changes in the price of gasoline.[8]

THE AGING POPULATION MAY DEMAND FEWER SMALL CARS

The mix of cars sold in the United States has changed substantially over the past two decades. Figure 1-5, in Chapter 1, shows the allocation of the fleet among light trucks and four automobile classes for model years 1975-1991. Clearly, there has been a shift away from larger cars, especially in the late 1970s, coinciding with the last of that decade's two oil shocks. One is tempted to infer from Figure 1-5 that there will not be a noticeable growth in sales of larger cars. The demography of the U.S. population suggests, however, that this trend may increase somewhat.

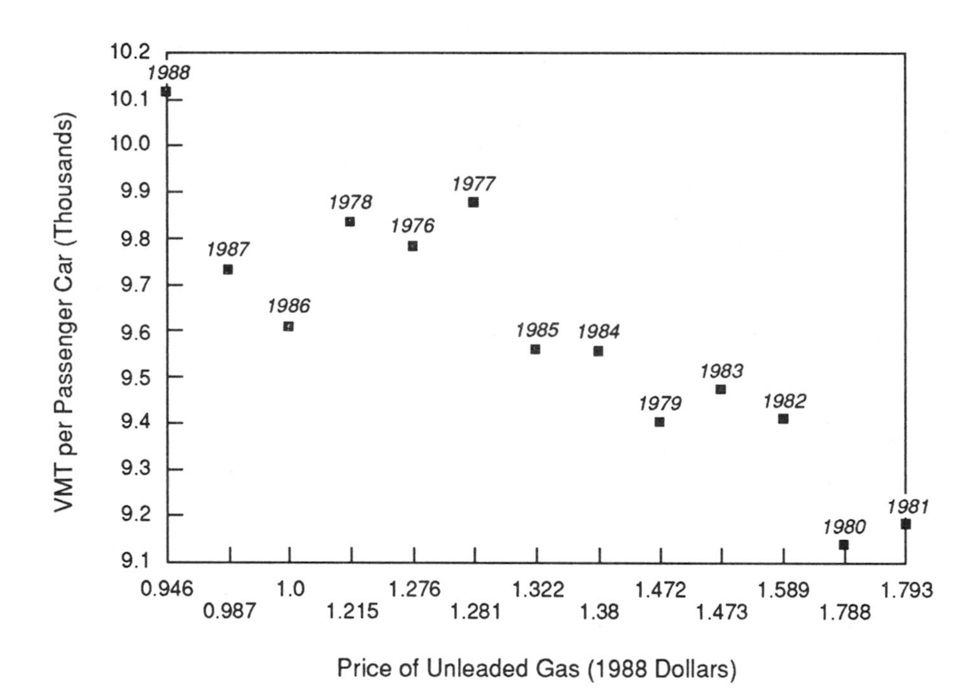

FIGURE 6-6 Annual vehicle miles traveled (VMT) per passenger car in relation to the price of gasoline (1976-1988).

SOURCE: Based on Oak Ridge National Laboratory (1991).

---

[8]There is considerable variation in estimates of the price sensitivity (elasticity) of vehicle miles traveled. Recent studies suggest very inelastic demand, in the range of -0.05 to -0.2 (Gately, 1990; Greene, in press), whereas earlier studies suggest a range of -0.3 to -0.9 (Dahl, 1986; Leone and Parkinson, 1990). (The price elasticity of miles traveled is the ratio of the percentage change in vehicle miles traveled to the percentage change in fuel price.)

Figure 6-7 shows how the population is expected to age over the coming decade and beyond. Based on Census Bureau projections, the number of people aged 35 to 54 years will increase substantially and the number aged 25 to 34 years will decline. These demographic changes may affect aggregate consumer preferences.

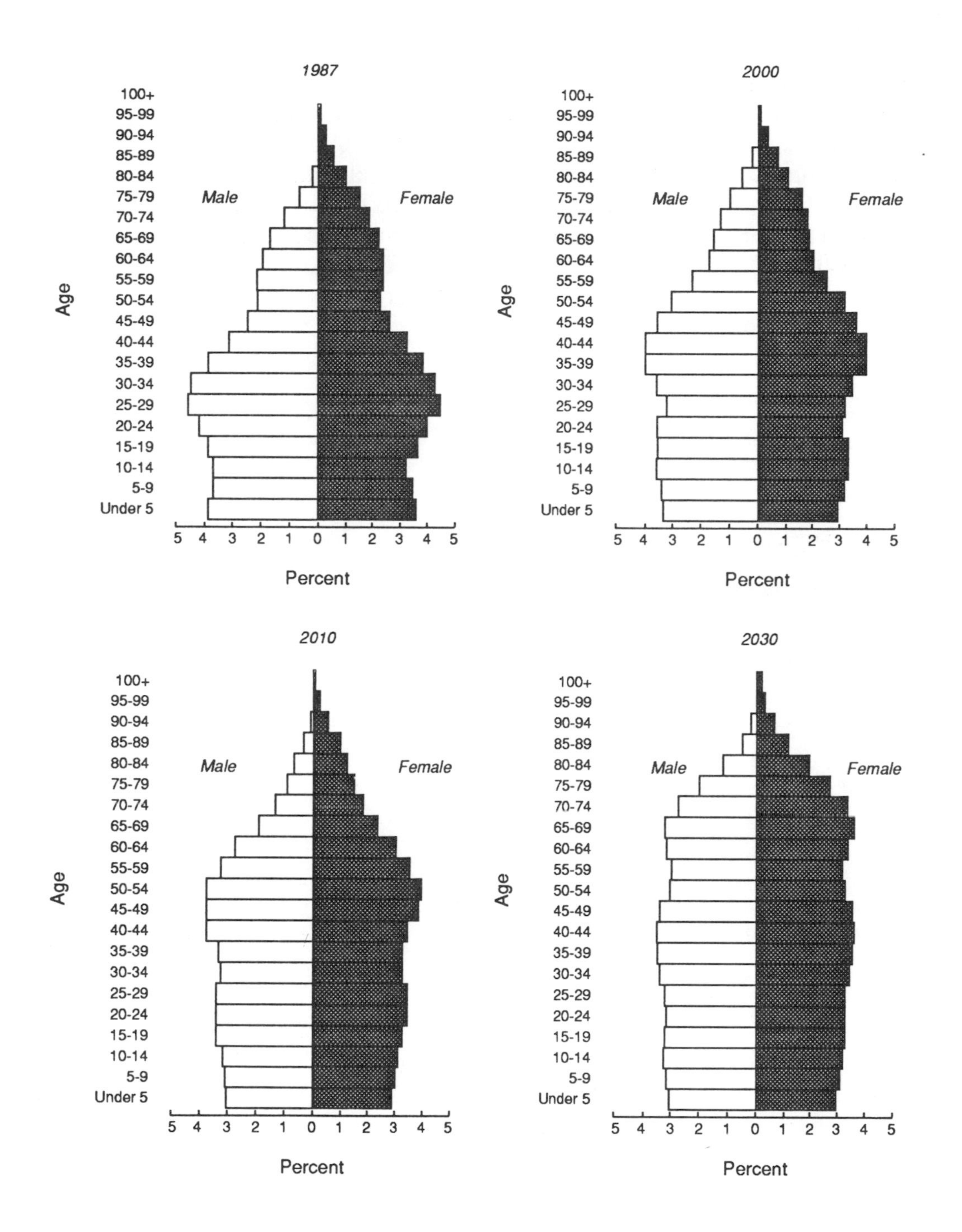

FIGURE 6-7 Age distribution of the U.S. population, 1987, 2000, 2010, 2030.

SOURCE: Bureau of the Census (1989).

Figure 6-8 shows current consumer tastes by age group. Young drivers (under age 34) clearly choose economy cars. As purchasers get older, they choose midsize cars. And older people (over age 55) clearly prefer large cars. The youngest people are typically childless and thus require room for only one or two people in a vehicle. As they move into their child-rearing years, they require a larger car to transport a family. As they age, they could go back to small cars, but historically they have not.[9]

Demographic trends and the age-based buying preferences of the current population suggest the possibility of an upsurge in demand for larger cars and a reduction in demand for smaller cars, all else being equal. This is not to suggest that current baby boomers will demand cars as large as their parents did, only that they may not move back to small cars. There is an alternative hypothesis, however--namely, that there has been a generational shift in tastes. One might argue that the current younger population is fundamentally different from the older population and will not make the move to large cars as they age. The demographic effects on the car market thus remain somewhat uncertain.

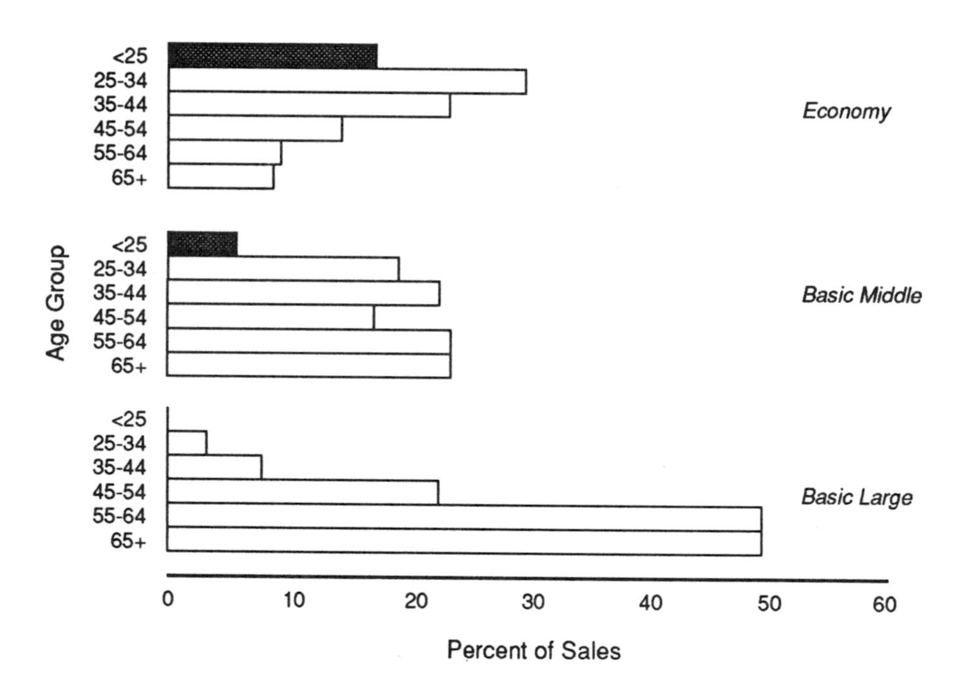

FIGURE 6-8 Car buyers by age group.

SOURCE: Presented to the Impacts Subgroup of the Committee on Fuel Economy of Automobiles and Light Trucks, September 16, 1991, by Chrysler Corporation based on data from Consumer Attitude Research (1991).

[9]There are several reasons why older people might prefer larger cars. Larger cars are generally safer. If age decreases the ability to drive, due to slowed reaction times and the like, then the likelihood of an accident will increase and a logical response is to purchase a safer car (at least for those in the larger car). A second factor is entry and egress, which are easier in large cars. A third factor is income--older people have more discretionary income and thus may be inclined to purchase a luxury vehicle.

SHIFTS IN AGGREGATE CONSUMER PREFERENCES

Two other trends in consumer preferences warrant comment. In the early 1980s there was a dramatic shift to light trucks--pickup trucks, vans, and utility vehicles. A major reason for this shift, at least initially, was that light trucks provide cheaper transportation than automobiles. Insurance was cheaper because it was based on historical use, which was largely commercial. Further, presumably because of their largely commercial use and less stringent environmental, emissions, and safety regulations, some categories of light trucks could be obtained in very basic versions without costly options. Other features of trucks also appealed to people, such as high seating.

The public interest in light trucks does bear on fuel economy standards. Although some uses of trucks require significant performance--for example, for commercial use and hauling purposes--many light trucks are simply used as automobiles. Further, the automotive industry created new types of trucks in the 1980s, particularly minivans, which are used in the same way as some categories of automobiles. But light trucks have been subject to a different fuel economy standard than automobiles. If this continues, and there is an increase in the use of these vehicles, a growing portion of the fleet will be subject to less stringent standards than automobiles, and fleet fuel economy attainments will be reduced.

Another trend over the past few decades relates to optional equipment. As Figure 6-3 indicated, consumers have chosen to purchase more automobile when they buy a new vehicle. How this translates into optional purchases of equipment is suggested in Table 6-1. Many of the options negatively affect fuel economy through increased weight and through energy consumption.

IMPACTS OF MANDATED FUEL ECONOMY ON THE CONSUMER

Probably the most subtle impacts of mandated fuel economy are those associated with the consumer. In the current marketplace, the consumer examines the features of new and used automobiles and the anticipated price of gasoline and decides whether to purchase an automobile and which characteristics to seek. Having acquired an automobile, the consumer decides how much to drive it. Fuel economy enters into the decision in terms of trading off an initial payment for fuel economy with a stream of payments for gasoline over the life of the car, taking into account the value of the car when the consumer disposes of it.[10]

---

[10]A car that has lower fuel economy will, all else being equal, have lower resale value. This lower resale value affects the decision to purchase a new car.

TABLE 6-1 Change in Optional Equipment for Selected Model Year Passenger Cars (in percentage of cars sold)

Options	1971	1980	1985	1990
Automatic transmission	91.1	82.7	86.3	88.4
Power steering	82.5	83.7	92.3	97.3
Power brakes (2- or 4-wheel disc)	59.3	64.6	95.6	92.4
Antilock brakes	--	--	--	7.6
Power seats	10.6	16.3	33.7	36.7
Air-conditioning (manual temperature control)	57.6	65.3	72.4	71.6
Air-conditioning (automatic temperature control)	5.8	7.7	14.8	20.3
Power windows	20.7	23.3	45.2	58.6
Driver airbag	--	--	--	29.7

SOURCE: MVMA (1991).

The propriety of government intervention in the market to require enhanced fuel economy might be guided in part by consideration of whether the consumer typically makes a rational choice in trading off fuel economy and the price of gasoline. While the evidence is far from complete, analyses of the prices of used cars (Kahn, 1986) suggest that the consumer does appropriately trade off the cost of fuel economy with the cost of gasoline in used car purchases. Difiglio et al. (1990) have calculated that the difference between such a rational consumer decision on fuel economy and a distorted one is on the order of 2 mpg.[11] As was noted earlier, however, total costs of owning and operating an automobile are relatively insensitive to fuel economy. Thus, because consumers must expend effort to determine the optimum in the trade-off of vehicle cost and fuel economy, error and even bias in consumer decisions may result.

A second important consideration concerns the public and private costs of gasoline consumption. While the consumer behaves on the basis of what he or she pays at the gas pump, the true societal costs of gasoline consumption may be higher. The costs of pollution and maintaining oil security are two costs directly attributable to gasoline

[11]In particular, Difiglio et al. (1990) assume that a rational decision involves amortizing a fuel economy investment over 10 years at a 10 percent real discount rate. They assume a distorted decision on fuel economy involves amortization over four years at the same discount rate.

consumption that do not appear in its price. These costs are difficult to quantify, but the security costs of oil have been estimated at anywhere from nothing to $3 per gallon (Broadman, 1986). If consumers do not pay the true cost of gasoline, they will undervalue fuel economy. Consequently, some mandated increase in fuel economy may enhance consumer well-being in the sense that the increased cost of fuel efficiency is justified by the societal and individual benefits of consuming less fuel.

It must be acknowledged, however, that raising the fuel economy standard above what the market would demand on its own has an impact on the consumer. For a particular car, a nonmarket-based increase in fuel economy has a cost.[12] As a result of standards, the consumer is faced with a menu of cars with increased prices (net of fuel costs) relative to the situation in which there is no mandated fuel economy standard. The consumer may make the same car choice before and after the price rise, in which case the additional cost of fuel economy comes directly out of the consumer's pocket. The more likely situation is that the consumer will look at the menu of choices and decide to take less of some characteristic, such as performance, size, or even safety. In this case, the consumer loss is somewhat less, but it is still a loss.

To summarize, as the mandated fuel economy standard is gradually raised above what would prevail without a standard, consumers first benefit, due to the fact that gasoline may be underpriced (for reasons stated above), and then they start to pay extra, either through higher car prices or through less desirable car characteristics. The full implications of the effects of fuel economy standards on the consumer are discussed in Chapter 9.

FINDINGS AND CONCLUSIONS

- The available information suggests that total annual sales of new cars will only grow moderately over the term of this study.

- The price of gasoline should enter into the consumer's decision to purchase a new vehicle. The rational consumer should trade off the additional cost of a new car that is associated with improved fuel economy against the resulting fuel savings. But gasoline price is not a very powerful lever for increasing the fuel economy of the new fleet. Because the total cost curve is relatively flat in the vicinity of the optimal fuel economy level, the consumer should not be expected to exert much effort to find the optimum, and the manufacturer has little incentive to take significant risks to produce cars that provide more fuel efficiency.

- As a result of the aging of the U.S. population, there could be an upsurge in demand for larger cars in the future, with adverse consequences for fuel economy. Such matters are not conclusively resolved by the available data, however; the baby-boomers may have different tastes than their parents.

---

[12] If there are costless ways of enhancing fuel economy, they should always be pursued.

- Since the early 1980s, there has been a major shift in consumer demand toward light trucks to serve the same purposes as automobiles. Because light trucks have lower fuel economy on average than automobiles, the trend has adversely affected fuel economy goals. Moreover, there has been increasing consumer demand for options that negatively affect fuel economy.

- Consumer decisions on the desirable fuel economy level in new cars are based on current and projected costs of gasoline. To the extent that this price is artificially low--that is, it does not reflect pollution, security, or other societal costs--the consumer will not choose a sufficient level of fuel economy when purchasing a new vehicle.

- Higher gasoline prices will cause people to drive less and thus will reduce gasoline consumption. Higher gasoline prices will also induce people to buy vehicles with greater fuel economy, although the effect may be modest.

REFERENCES

Broadman, H.J. 1986. The social cost of imported oil. *Energy Policy* June:242-252.

Bureau of the Census. 1989. *Projections of the Population of the United States, by Age, Sex, and Race: 1988 to 2080*. Current Population Reports, Series P-25, No. 1014. Washington, D.C.: U.S. Government Printing Office.

Bureau of the Census. 1991. *Statistical Abstract of the United States*. 111th ed. Washington, D.C.: U.S. Government Printing Office.

Consumer Attitude Research (C.A.R). 1991. *C.A.R. Report*. 2nd qtr. Birmingham, Mich.

Dahl, C. 1986. Gasoline demand survey. *Energy Journal* (7):67-82.

Difiglio, C., K.G. Duleep, and D. Greene. 1990. Cost effectiveness of future fuel economy improvements. *Energy Journal* 11(1):65-86.

Gately, D. 1990. The U.S. demand for highway transportation and motor fuel. *Energy Journal* 11(3):59-73.

Greene, D. In press. Vehicle use and fuel economy: How big is the rebound effect? *Energy Journal*, forthcoming.

Heavenrich, R.M., J.D. Murrell, and K.H. Hellman. 1991. *Light-duty Automotive Technology and Fuel Economy Trends Through 1991*. Control Technology Applications Branch, EPA/AA/CTAB/91-02. Ann Arbor, Mich.: U.S. Environmental Protection Agency.

Kahn, J.A. 1986. Gasoline prices and the used automobile market: A rational expectations asset price approach. *Quarterly Journal of Economics* May:323-339.

Leone, R.A., and T.W. Parkinson. 1990. *Conserving Energy: Is There a Better Way? A Study of Corporate Average Fuel Economy Regulation*. Washington, D.C.: Association of International Automobile Manufacturers.

Motor Vehicle Manufacturers Association (MVMA). 1991. *Facts & Figures '91*. Washington, D.C.

Oak Ridge National Laboratory. 1991. *Transportation Energy Data Book*. 11th ed. ORNL-6649. Oak Ridge, Tenn.

7

FUEL ECONOMY PROJECTIONS

This chapter describes the methods, assumptions, and results of three approaches the committee used to make projections of potential fuel economy levels. The results are "raw material" for Chapter 8, in which the committee reports its judgments of achievable fuel economy levels.

PROJECTING FUEL ECONOMY LEVELS

Previous Efforts

Projecting future vehicle fuel economy is a risky business. The recent history of such endeavors makes it clear that the chances of being very wrong are very high. In the late 1970s and early 1980s, a number of studies attempted to project fuel economy levels for automobiles and light trucks through 1990. Most of the studies overestimated fleet fuel economy levels by a substantial amount. Estimates for 1990 passenger cars ranged from approximately 30 to 40 miles per gallon (mpg), but the actual fuel economy level was 28 mpg; estimates for light trucks ranged from 20 to 30 mpg, compared with the actual 20 mpg (U.S. Department of Transportation, 1991). The large differences arose mainly because the analysts anticipated continually rising gasoline prices, whereas in the late 1980s gasoline prices were considerably lower in real terms than in the late 1970s.

More recent projections of fuel economy into the early twenty-first century for passenger cars are summarized in Table 7-1. Projections of the fuel economy of the new-car fleet for model year (MY) 2001 range from about 29 to 45 mpg. The wide variation in forecasts is apparently due to such matters as differences in assumptions about what is acceptable to consumers and what is technically possible.

Part of the explanation for the divergence of past projections from subsequent history and for the considerable variation among contemporary projections is that the

TABLE 7-1 Recent Projections of the Average Fuel Economy of Future New Passenger-Car Fleets

Source[a]	Fleet-Average Fuel Economy (mpg) MY 1995-1996	MY 2000-2001	MY 2005-2006	MY 2010
ACEEE[b]	---	45	---	---
Berger et al.[c]	29.5-30.3	30.3-32.8	---	---
Bryan bill[d]	34.3	39.5	---	---
Chrysler				
Engineering assessment[e]	30.1	30.9	---	---
Max technology	---	34.5	---	---
EEA[f]				
Product plan	28.3	32	---	---
Max technology	29.1	36	---	---
Risk level 1	---	---	---	44.8
Risk level 2	---	---	---	54.9
Risk level 3	---	---	---	74.1
DOE[g]				
Product plan	32.9	36.2	---	---
Cost effective	---	38.6	---	---
Max technology	---	42.0	---	---
Industry[h]	30.2	30.6	---	---
Johnston bill[i]	29.9	33.6	37	---
Ledbetter & Ross[j]	---	40.1-43.8	---	---
Lovins[k]	(no specific time frame)		(70-100)	
OTA[l]				
1995	29.2-30.0	---	---	---
2001	---	32.9-38.2	---	---
2005	---	---	37.1-38.1	---
2010	---	---	---	45-55
SRI[m]	28.1	28.7-30	---	---

(See table notes on next page.)

[a] Much of the data in this chart is based on Office of Technology Assessment (OTA, 1991).

[b] Analysts at the American Council for an Energy-Efficient Economy argue that a year-2000 goal of 45 mpg is technologically attainable and cost-effective at today's gasoline prices (Chandler et al., 1988).

[c] Berger et al. (1990) conducted statistical analysis for fuel economy as a function of vehicle attributes: 4.92 to 7.98 percent improvement from 1987 to 1995 and 9.62 to 16.6 percent improvement for 1995-2001.

[d] Sen. Richard Bryan's bill (S.279): 20 percent over 1988 by 1996; 40 percent by 2001.

[e] Bussmann (1990) indicates 7.1 percent gain from 1989-1995 and 2.8 percent from 1995-2000. Under maximum technology case, 14.6 percent gain from 1995 to 2000 (see OTA, 1991:Table 7-2).

[f] Energy and Environmental Analysis, Inc. Estimates are for the domestic fleet only: risk level 1 refers to technologies most agree are likely to be commercialized; level 2 is technology about which opinion is sharply divided; level 3 is technology considered esoteric by most, but within the realm of possibility (OTA, 1991:58).

[g] U.S. Department of Energy: product plan, 17.1 percent increase from 1987-1995, and 9.9 percent increase from 1995-2000. Cost-effective assumption, 17.2 percent increase from 1995-2000; maximum technology, 27.6 percent from 1995-2000 (see OTA, 1991:Table 7-1).

[h] Estimates presented by the automotive companies at the committee's July 1991 workshop (see Appendix F) were on the order of a 10 percent increase from 1990 to 2001. Toyota indicated an 11 percent increase by 2001; Chrysler indicated a 17 percent increase for small cars, 19 percent for midsize, and 6 percent for large cars from 1990-2001; Honda indicated 10 to 11 percent for 1990-2001 (without considering emissions and safety requirements, 14 to 15 percent); General Motors indicated 1 to 2 mpg above 24 mpg would be cost-effective for a gasoline price of $1.34/gallon in 2000 for the Buick Park Avenue; Nissan reported 10.6 percent from 1990-2000; Mitsubishi reported 8 percent. OTA (1991:Table 7-3) reports a 7.6 percent improvement from 1989 to 1995 for the domestic industry.

[i] Sen. Bennett Johnston's bill (S.1220): 7.5 percent increase from 1990-1996; 20.9 percent from 1990-2001; 31.8 percent from 1990-2006. The 1996 standard involves nearly full market penetration of fuel-saving technologies and substantial penetration by the two-stroke engine. Light-truck corporate average fuel economy (CAFE) levels are 22.0, 24.0, and 26.6 mpg for 1996, 2000, and 2006, respectively, for the combined domestic and import fleets.

[j] Ledbetter and Ross (1990) use a variation on Energy and Environmental Analysis, Inc.'s analysis for OTA (1991) that is based on more optimistic assumptions.

[k] Amory Lovins at the committee's July 1991 workshop (see Appendix F) indicated large increases with reduced costs but no specific time frame.

[l] OTA (1991) estimates range from optimistic to pessimistic values depending on the scenario envisioned. From 1995 to 2001, the scenarios range from a fairly optimistic "business as usual" at the low end to a regulatory, technology-forcing scenario at the high end. From 2005 on, regulation-driven scenarios are envisioned both for the low and high estimates in 2006 and 2010.

[m] SRI (1991).

projections have entirely different foundations. Some projections reflect what the analysts would prefer to have happen and vary substantially in the degree of recognition of practical constraints or, alternatively, future opportunities. Other projections represent the analysts' best estimates of what will happen, not necessarily what they prefer to have happen. The analysts may attempt to account for specific anticipated changes that bear on fuel economy projections, or they may assume that change of some unspecified nature will occur with the passage of time. Still other projections attempt to establish the limits of what is possible, with little or no reference to the degree to which the possible is feasible. Finally, some projections attempt to take into account the possibility of radical change in technology, in values, or in the political, economic, or natural conditions that impinge on fuel economy. In light of these facts, the differences among the various estimates are not surprising.

Overview of the Projection Methods

The committee used three types of quantitative projections: historical trend projection; "best-in-class" (BIC) analysis; and the technology-penetration, or "shopping cart," approach.

Historical trend projections are based on the assumption that the past trend in some measure of overall performance can be extended reliably into the future. This study projects recent experience with the fuel economy of the new vehicle fleet to MY 2006. It also projects a weight-adjusted measure of the technical efficiency of the new vehicle fleet, namely ton-miles per gallon (ton-mpg), over the same period.

Best-in-class analysis is a hybrid approach. This type of analysis assumes that the fuel economy of the typical new vehicle at some unspecified future date will be equal to that of an exemplary vehicle or group of vehicles already in the fleet.

The technology-penetration, or shopping cart, approach attempts to locate the limits of possibility, subject to certain constraints, by assuming that specific, well-characterized, fuel-saving technologies will be adopted in a larger proportion of all vehicles than they are today. Unlike the trend-projection and BIC methods, the shopping cart approach provides the basis for an explicit projection of costs.

Assumptions Common to All Projections

Several assumptions underlie all of the fuel economy projections in this study. First, such amenities as air-conditioning, automatic transmissions, high ride quality, and brisk acceleration are assumed to continue at current levels. For example, horsepower-to-weight ratios (a performance measure) are not changed in most of the projections from the base MY 1990 levels;[1] technologies having potentially serious customer-acceptance problems, such as engine-off at idle, are excluded from the shopping cart

---

[1]Some estimates are reported of the extent to which fuel economy could be improved by allowing performance levels to be reduced.

analysis; and MY 1990 market share levels are kept constant for certain technologies that have important implications for consumers.[2]

A second assumption is that there will be a "normal" rate of technological progress in fuel economy improvement between now and 2006. Higher levels of fuel economy will be possible in the future than are possible today, at reasonable costs, but radical breakthroughs in technology are assumed not to occur.

Third, it is assumed that the means used to improve fuel economy will be consistent with compliance with the Tier I emissions standards of the 1990 amendments to the Clean Air Act, as discussed in Chapter 4. This constraint is reflected in the assumption that there will be an overall fuel economy penalty of 1 percent due to the added weight of emissions-control equipment. No other impact on vehicle performance is assumed to result from compliance with new emissions standards. It is further assumed that automobile makers will conform with expected future safety standards, as discussed in Chapter 3, with a consequent fuel economy penalty of 2 percent for the added weight likely to be necessary to meet those requirements.

Fourth, it is assumed that there will be no shifts of consumer preferences among the size classes or from automobiles to light trucks. The committee uses the Environmental Protection Agency's (EPA's) vehicle classification system of nine passenger-car and six light-truck classes. Projections were made for the four largest selling classes of cars (subcompacts, compacts, midsize, and large) and trucks (small pickups, large pickups, small vans, and small utility vehicles). These classes account for 92 and 88 percent of car and light-truck sales, respectively (Heavenrich et al., 1991).[3]

HISTORICAL TREND PROJECTIONS

Although extrapolation of past trends may not always accurately reveal the future, there is no reason to believe that extrapolation of fuel economy trends is inevitably unreliable. Indeed, trend projection has the advantage of being inherently temporal and, thus, able to suggest how rapidly improvements might be achieved. The method provides a crude but revealing indicator of possible fuel economy gains.

As discussed in Chapter 1, fuel economy gains averaging 1 to 1.5 mpg per year were achieved during the 1970s as a result of vehicle downsizing and downweighting associated with the conversion of most of the fleet to front-wheel drive, the development of the three-way catalyst, and other technological improvements. During the mid-1980s, vehicle weights stabilized, and technological improvements resulted in

---

[2]These include, for example, market shares of automatic and manual transmissions and shares of 4-wheel drive in light trucks. In reducing the number of cylinders in engines, engines are changed from eight to six cylinders and from six to four, but not from eight to four.

[3]The omitted car classes are two-seaters, minicompacts, and small, midsize, and large wagons. Light trucks omitted are large vans and large utility vehicles.

a slower rate of fuel economy increases (0.4 to 0.5 mpg per year). From the late 1980s to the present, acceleration performance was enhanced and vehicle weights increased slightly, which resulted in a small reduction in fleet fuel economy.

The historical trend projections of fuel economy were made by linear extrapolation of trends in two measures--mpg and ton-mpg--using the method of linear least squares to fit trend lines for the 16-year base period, 1975-1991, as illustrated in Figures 7-1 and 7-2.[4] Other projection lines could be obtained by fitting subsets of the historical data. For example, during the early part of the base period, rising fuel prices, anticipation of growing fuel shortages, and the need to meet future fuel economy requirements motivated rapid improvement of fuel economy. Projecting the rate of progress during that period would lead to very high fuel economy forecasts for MY 2006.[5] On the other hand, during later parts of the period, compliance with CAFE requirements may have been the only major reason for manufacturers to pay attention to fuel economy, and progress slowed. Thus, projecting the experience of the past decade would lead to a forecast of quite small improvements. The committee recognizes that fleet-average fuel economy closely followed the mandated fuel economy standards over the 1978-1991 period and does not fit a straight line particularly well. Nonetheless, in the absence of any obviously preferable choice, the committee projected the straight lines over the entire period to estimate future fuel economy progress.

The trend-based extrapolation results for cars and light trucks are summarized in Tables 7-2 and 7-3. The mpg-based MY 2006 projections for the new car and truck fleets were obtained from the linear extrapolations beyond MY 1991. (Each MY 2006 value was reduced by 3 percent to account for the fuel economy penalty associated with meeting anticipated safety and emissions requirements.) The mpg-based MY 2001 projections for the new car and truck fleets were then obtained by linear interpolation between the actual MY 1991 level and the adjusted MY 2006 projections. The ton-mpg projections for all cars and all trucks were obtained similarly, assuming that the average weight of all cars and trucks in MY 2001 and MY 2006 remain at the reported MY 1991 values. Projections for each size class of cars and trucks for MY 2001 and MY 2006 were then obtained by applying the percentage increases that apply to each fleet to the reported MY 1991 values for each size class.

---

[4]The mpg measure is the fleet-average fuel economy for the new car and truck fleets evaluated using the EPA Federal Test Procedure. The ton-mpg measure is the fleet-average fuel economy of cars and trucks (in mpg) multiplied by vehicle weight in tons. Data are taken from Heavenrich et al. (1991). The ton-mpg measure partially adjusts for major changes in weight and helps focus on improvements in engine and drivetrain efficiency, aerodynamics, and rolling resistance. The ton-mpg method is used here on a purely empirical basis. It accounts only approximately for changes in vehicle weight in that the fractional change in fuel consumption is equal to roughly 0.7 of the fractional change in the vehicle weight. Weight-adjusted fuel economy as a measure of technical efficiency has been used by many analysts (see, for example, Samples and Wiquist, 1978).

[5]Similarly, the gross overestimates for 1990 reported above are consistent with projecting the 1975-1980 experience to 1990.

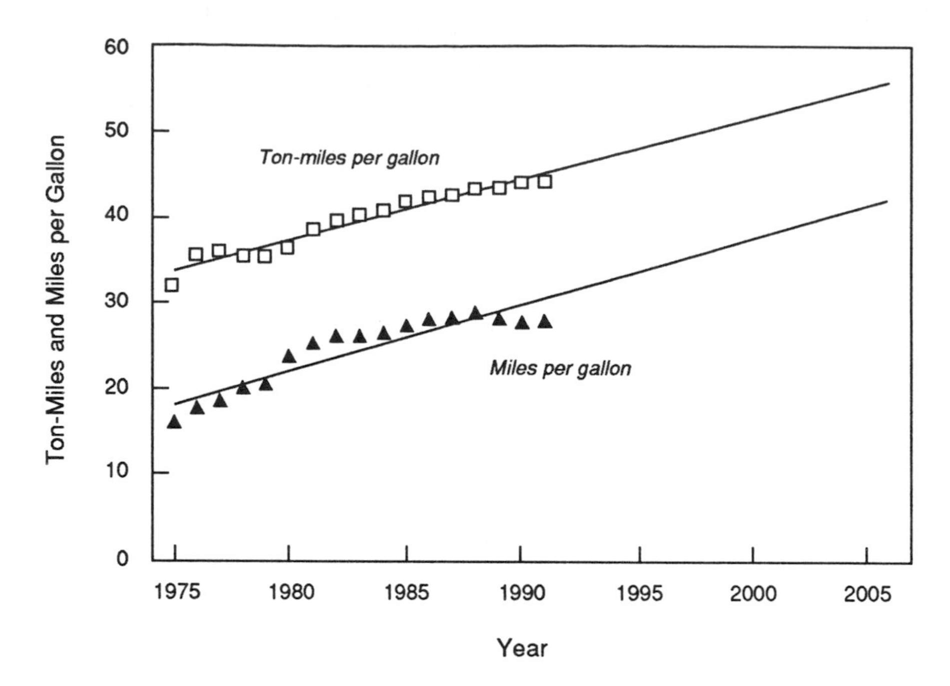

FIGURE 7-1 Fuel economy trends for the new passenger-car fleet.

SOURCE: Data from Heavenrich et al. (1991). Projections by the committee.

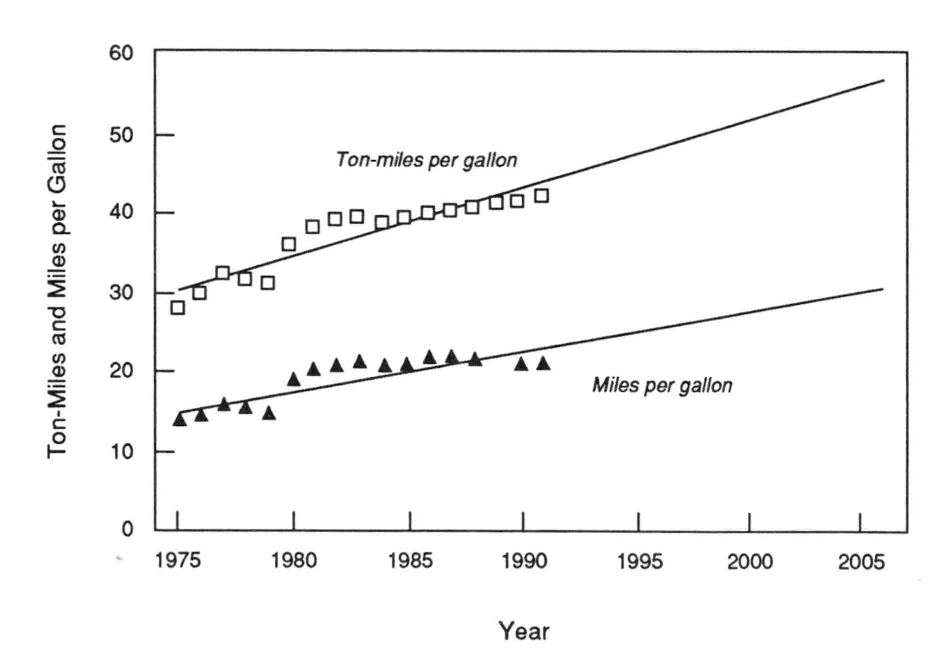

FIGURE 7-2 Fuel economy trends for the new light-truck fleet.

SOURCE: Data from Heavenrich et al. (1991). Projections by the committee.

TABLE 7-2 Trend Projections of Fuel Economy of New Passenger Cars

Size Class and Model Year	Fuel Economy (mpg) Based on MPG Trends for 1975-1991	Based on Ton-MPG Trends for 1975-1991
Car fleet		
1991	27.8	27.8
2001	34.9	32.3
2006	38.5	34.5
Subcompact cars		
1991	31.2	31.2
2001	39.2	36.2
2006	43.2	38.7
Compact cars		
1991	29.2	29.2
2001	36.7	33.9
2006	40.4	36.2
Midsize cars		
1991	25.8	25.8
2001	32.4	29.9
2006	35.7	32.0
Large cars		
1991	23.7	23.7
2001	29.8	27.5
2006	32.8	29.4

NOTE: Projections for MY 2006 have been adjusted downward by 3 percent to account for the fuel economy impact of safety and emissions (Tier I) standards. Data for 1991 are from Heavenrich et al. (1991). These projections are considered as input to the committee's estimates reported in Chapter 8.

TABLE 7-3 Trend Projections of Fuel Economy of New Light Trucks

Size Class and Model Year	Fuel Economy (mpg) Based on MPG Trends for 1975-1991	Based on Ton-MPG Trends for 1975-1991
Light truck fleet		
1991	20.8	20.8
2001	25.2	24.8
2006	27.5	26.8
Small utility trucks		
1991	21.2	21.2
2001	25.7	25.2
2006	28.0	27.3
Small vans		
1991	22.8	22.8
2001	27.7	27.2
2006	30.1	29.3
Small pickups		
1991	25.2	25.2
2001	30.6	30.0
2006	33.3	32.4
Large pickups		
1991	19.5	19.5
2001	23.7	23.2
2006	25.8	25.1

NOTE: Projections for MY 2006 have been adjusted downward by 3 percent to account for the fuel economy impact of safety and emissions (Tier I) standards. Data for 1991 are from Heavenrich et al. (1991). These projections are considered as input to the committee's estimates reported in Chapter 8.

The mpg-based projections indicate that the MY 2006 *fleet-average* fuel economy might increase by about 38 percent for passenger cars and by about 32 percent for light trucks above MY 1991. The ton-mpg-based projections indicate that the MY 2006 fleet-average fuel economy might increase by about 24 percent for cars and 29 percent for trucks above MY 1991. For passenger cars, the fuel economy projections based on mpg are higher than those based on ton-mpg, in part because future weight reductions are assumed not to occur in the ton-mpg projections, whereas weight reduction is implicit in the mpg projections. The ton-mpg projections would have been higher had future weight reductions been taken into account. For light trucks, projections based on the two methods are nearly identical because weight reduction has not historically been a significant factor in their fuel economy improvement.

BEST-IN-CLASS (BIC) PROJECTIONS

The BIC method is based on the assumption that the average fuel economy of each vehicle class could eventually reach the level of the best vehicle, or several top vehicles, currently in that class. The method implicitly assumes that the cost of achieving the BIC fuel economy level for an entire class would be no greater than the costs of lost consumer satisfaction and manufacturers' profits as a result of restricting consumer choices to the BIC vehicle(s) or ones of similar design. Of course, if technology were to improve, the fuel economy performance of the BIC vehicle(s) might be achieved by vehicles with a wider range of amenities in the future.

There is an implicit time period for achieving the BIC fuel economy level: the time it would take to redesign, produce, and sell vehicles in each class such that the class-average fuel economy would be equal to the fuel economy of the current BIC vehicles. Platform designs, as noted earlier, are essentially established through MY 1995, but the BIC levels estimated in this report should, according to the discussion in Chapter 5, be achievable by manufacturers by MY 2006 without undue disruption to normal product cycles.

The EPA annually carries out BIC fuel economy analyses for the fleet and each weight class (Heavenrich et al., 1991). It ranks individual configurations (a car line with a specific engine and transmission combination) without regard to sales volume, and it identifies the BIC vehicle, the five BIC vehicles, and the top dozen vehicles within each weight class (the weight classes differ from the volume-based size classes used throughout this report).[6]

---

[6]For MY 1990 passenger cars, the EPA estimated a 24 percent improvement in the fleet-average fuel economy, to 34.4 mpg, assuming that the fuel economy of the average vehicle would equal that of the single BIC and maintaining the actual 1990 sales distribution by weight class. The EPA estimated a 20 percent improvement if the average fuel economy reached that of the top dozen BIC vehicles in each class. For trucks, a BIC analysis based on the single BIC vehicle in each class suggested a 23 percent improvement in fleet-average fuel economy, to 25.9 mpg.

The EPA also ranks *all* configurations according to their fuel economy and identifies the average fuel economy of various percentile groups. Figure 7-3 shows the trends in fuel economy for each percentile group. It shows that the average passenger car in MY 1986 through MY 1991 achieved a fuel economy (28 mpg) about equal to that of the ninety-fifth percentile car in MY 1975. Roughly 10 years were required to accomplish that change. In MY 1991, the ninety-fifth percentile passenger car had a fuel economy of about 45 mpg. This chart makes clear that the very top-ranked cars (in mpg) have become decreasingly typical of the rest of the fleet over time.

In this study the BIC vehicle(s) are chosen using an approach different from EPA's. Vehicles are defined by car line, not by configuration, and classed by interior volume (cars) or size and function (trucks), not by weight.[7] The BIC vehicles are those in the top rank that together account for at least 5 percent of class sales. The BIC fuel economy for each class is calculated as the sales-weighted harmonic average fuel economy (see Chapter 9) of the BIC car lines, reduced by 3 percent for the effects of safety and emissions standards.

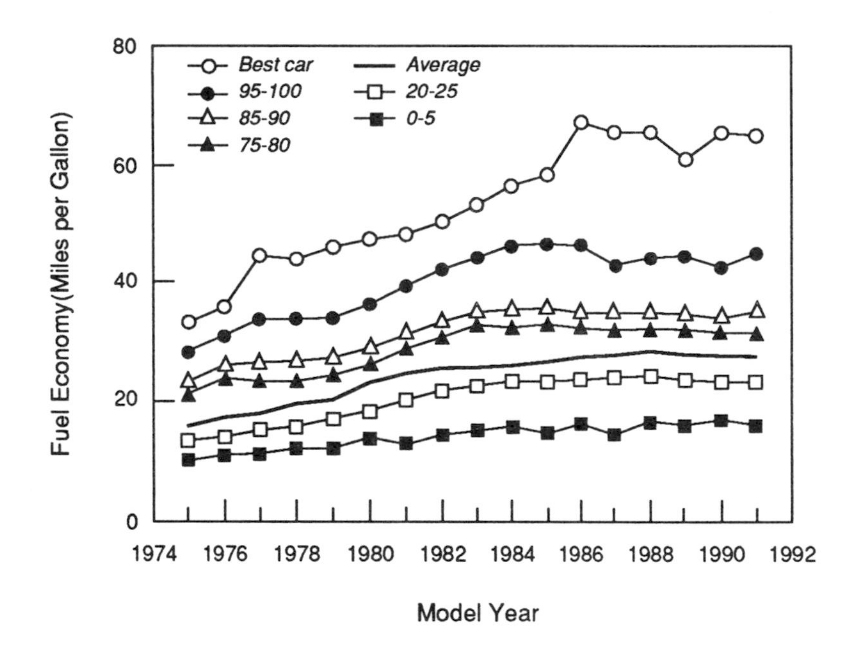

FIGURE 7-3 New passenger-car fuel economy by percentile group.

SOURCE: Heavenrich et al. (1991).

---

[7]A car line typically includes several drivetrain variations, a reflection of consumer choices of engines and transmissions. Thus, in conformance with actual purchase patterns, these BIC estimates are based on car lines that include a large proportion of automatic transmissions, for example, whereas EPA's method identifies as "BIC" a group of cars equipped almost exclusively with manual transmissions.

Tables 7-4 and 7-5 show the results of the BIC analysis based on MY 1990 vehicles. For passenger cars, this analysis yields BIC fuel economy levels that are 50 percent above the current class average for subcompacts, about 13 percent for compacts, 8 percent for midsize cars, and 3 percent for large cars. There is a wide range of fuel economy values among the BIC vehicles in the subcompact class, a market segment in which fuel economy is considered an especially important attribute by many car buyers. Some manufacturers introduce cars in this class that are intended to compete for the title of the most fuel-efficient car sold in the United States. As a consequence of that competition, the most efficient subcompact car tends to have fewer of the other attributes most valued by consumers, even for its size class. As a result, the BIC analysis may exaggerate the potential fuel economy improvement for smaller cars. In contrast, in the large car class there is little variation in fuel economy among the BIC car lines. To the extent that there is no competition for the most efficient car in the other size classes, the BIC analysis may underestimate potential fuel economy gains.

The BIC fuel economy levels for the four light-truck classes shown in Table 7-5 represent modest gains for the small pickup and compact van classes (on the order of 5 percent), but substantial gains for large pickups and small utility vehicles (38 percent and 40 percent, respectively). This result best illustrates a weakness of the BIC approach applied at the vehicle-class level--it is sensitive to a few extreme cases and to the precise definitions of the classes. For example, the BIC large pickup truck is the Toyota, which has a fuel economy rating of 27.1 mpg, whereas the top-ranked domestically manufactured large pickup is the Dodge Power Ram 50, which has a fuel economy rating of 23.4 mpg, 15 percent lower. However, large pickups manufactured in the United States are considerably larger than the imported large pickups that rank at the top of the fuel economy scale, even though the definition of the large pickup class is broad enough to include both. Thus, the committee believes that the BIC method, as applied here to light trucks, gives only a general indication of fuel economy potential by class.

TECHNOLOGY-PENETRATION OR SHOPPING CART PROJECTIONS

Method and Assumptions

The technology-penetration projection method is based on an explicit consideration of the potential fuel economy contributions of specific, well-established technologies. It is referred to here as the "shopping cart" approach because it allows selection, one-by-one, of a set of fuel economy technologies, followed by a calculation of their contributions to vehicle fuel economy and cost. The shopping cart approach, as used by the committee, assumes that each technology will achieve its maximum possible market penetration--up to 100 percent where appropriate--and that each technology will be fully deployed before the next most cost-effective technology's market share increases above its base line.

TABLE 7-4 Best-In-Class (BIC) Fuel Economy, MY 1990 Passenger Cars

Size Class and Car Line	Market Share In Size Class	Fuel Economy (mpg)
Subcompact		
Chevrolet Sprint/Geo Metro	0.026	52.6
Geo Metro (domestic)	0.013	52.0
Daihatsu	0.006	43.1
Subaru Justy	0.008	39.3
BIC	0.052	47.2
Compact		
Isuzu Stylus	0.000[a]	35.4
Ford Escort	0.088	34.4
BIC	0.088	33.4
Midsize		
Dodge Aries/Plymouth Reliant	0.005	30.0
Mazda 626	0.011	29.6
Mazda MX6	0.000[b]	29.1
Chevrolet Corsica	0.068	28.3
BIC	0.084	27.7
Large		
Buick Electra/Park Ave.	0.044	25.0
Buick LaSabre	0.119	25.0[c]
BIC	0.163	24.3

NOTE: The car lines listed are those in the top rank for that class that together account for at least 5 percent of class sales. The BIC fuel economy for each class is the sales-weighted harmonic average fuel economy of these car lines adjusted downward by 3 percent to account for safety and emissions (Tier I) standards.

[a] 326 cars sold.
[b] 249 cars sold.
[c] The next three large cars in rank order also rated 25.0 mpg: Oldsmobile 88, Oldsmobile 98, and Pontiac Bonneville.

SOURCE: Data are from Williams and Hu (1991): Tables 16-19.

TABLE 7-5 Best-In-Class Fuel Economy, MY 1990 Light Trucks

Size Class and Car Line	Market Share in Size Class	Fuel Economy (mpg)
Small pickups		
Nissan pickup (domestic)	0.144	27.9
BIC	0.144	27.1
Small vans		
Plymouth/Dodge Vista	0.006	27.3
Toyota cargo van	0.001	25.1
Toyota van/Previa	0.034	25.0
Dodge Caravan	0.215	25.0
BIC	0.256	24.3
Large pickups		
Toyota pickup	0.067	27.1
BIC	0.067	26.3
Small utility vehicles		
Suzuki Samurai	0.007	33.5
Suzuki Sidekick	0.014	30.7
Geo Tracker	0.001	30.7
Geo Tracker (domestic)	0.040	30.2
BIC	0.062	29.8

NOTE: The trucks listed are those in the top rank for that class that together account for at least 5 percent of class sales. The BIC fuel economy for each class is the sales-weighted harmonic average fuel economy of these vehicle lines, adjusted downward by 3 percent to account for safety and emissions (Tier I) standards.

SOURCE: Data are from Williams and Hu (1991): Tables 29-31 and 33.

The committee used the shopping cart approach to construct curves relating fuel economy improvement to cost. In doing so, it assumed that the individual technologies are adopted in the order of decreasing cost-effectiveness (mpg per dollar) and that their incremental costs and fuel economy contributions are independent of market share.[8] The base year is MY 1990. Since the method uses only proven technologies, it assumes that the technologies can be implemented by manufacturers in the normal process of replacing manufacturing equipment, that is, in 15 years or less. Thus, estimates can be made of the costs of improving fuel economy to any particular level, so that curves of cumulative cost versus fuel economy improvement can be generated. Although these cost curves must be interpreted with care, for reasons pointed out below, they are quite useful for indicating the general tendencies for costs and risk to increase as higher levels of fuel economy are pursued.

As each technology is added to vehicles in a size class, the average increase in fuel economy for a car or light truck in the class is computed by multiplying an estimate of the percentage improvement in fuel economy achievable in a single vehicle by the change in market share for the technology from the MY 1990 level to the maximum possible for the class, and then multiplying the result by the base-year average fuel economy for the class. The average increase in cost for the size class due to increased use of each technology is computed by multiplying the average cost increase by the average market share change, both in the class. For the set of technologies applicable to a size class, summing the average fuel economy gains and average costs yields estimates of aggregate fuel economy gains and costs for the size class. The average gain in fuel economy is then added to the average fuel economy for the size class in the base year (3 percent having been subtracted from the base-year fuel economy to account for the effects of safety and emissions technology) to obtain an estimate of the fuel economy potential of the size class. (The method is detailed in Appendix E.)

The shopping cart estimates involve several other assumptions:

- Only proven technologies are included. No allowance was explicitly made for the development of new technologies or for the refinement of existing ones. This conservatism gives some assurance that the fuel economy gain could actually be achieved.

- For the main shopping cart analysis, vehicle performance was held constant at MY 1990 levels. (The effect of reducing horsepower/weight ratios to those of new MY 1987 vehicles was also examined to explore the kinds of trade-offs that might be made.)

---

[8]To do this, it was necessary to combine technologies in a logical way. Thus, all transmission changes were combined into a single category, as were all fuel system changes. In addition, redesigning engines to use fewer cylinders was combined with application of four valves per cylinder, overhead camshaft, and variable valve timing because the committee believes that the combined use of these technologies would be necessary to make an engine with a smaller number of cylinders acceptable to consumers.

- The analysis included downsizing associated with conversion of most of the remaining rear-wheel drive vehicles to front-wheel drive, as well as a 10 percent weight reduction through materials substitution.[9]

- The maximum market shares for each technology for each vehicle type were assumed to be 100 percent unless there were compelling reasons to limit market penetration.[10]

- Positive and negative synergistic effects among technologies were neglected, and thus, individual percentage improvements for different technologies were assumed to be additive.[11]

Data

The shopping cart algorithm requires several key inputs. The MY 1990 base-year composite fuel economy rating used for each car and light-truck class is that reported by Heavenrich et al. (1991). Data on the characteristics, fuel economy contributions, costs, and market shares of each technology were obtained from a variety of sources. The principal sources are reports compiled by SRI International for the Motor Vehicle Manufacturers Association (SRI, 1991) and by Energy and Environmental Analysis (EEA), Inc., a U.S. Department of Energy contractor (EEA, 1991a,b, and personal communication, October 2, 1991). Details of the data sources are provided in Appendix B. As discussed below and in Appendix B, there are substantial differences among these reports, especially regarding costs.[12]

The committee compiled a list of available fuel economy technologies applicable to conventional automobiles and light trucks based on information provided by automobile manufacturers and others who made presentations to the committee (see

---

[9]Twelve percent of the passenger cars and 25 percent of the light-truck fleet are downsized by conversion to front-wheel drive. The fleet-average weight reduction associated with this downsizing is about 60 pounds per car and 70 pounds per truck.

[10]The number of cylinders is reduced by no more than two to avoid consumer objections to noise and vibration. Increases in the market share of manual transmissions above the base year are not considered because consumer reaction would almost surely be strongly negative. Similarly, the base-year market share of 4-wheel drive trucks is not reduced. Continuously variable transmissions (CVT) are used only in subcompacts and 10 percent of compact cars because current technology does not permit CVTs to be used with higher torque engines. Diesel engines are used only in larger light trucks because of the problems of meeting passenger-car emissions standards.

[11]This assumption only approximates a more complex reality. For example, the 4-valve engine extends the flexibility of variable valve control technology and has beneficial effects on combustion chamber design. However, the committee believes that accounting for such interactions would not substantially affect the results of the analysis.

[12]Because it could find no estimates for light trucks, the committee assumed that the costs for technologies are the same for passenger cars and light trucks.

Table 2-2). The list includes the most important fuel economy technologies that, in the committee's view, satisfy the criteria of being in mass production and not degrading consumer satisfaction. Appendix B includes descriptions of the selected technologies and estimates of their fuel economy improvement potentials and costs. The list is generally consistent with available published studies, such as that of OTA (1991), and it is nearly identical to the list provided to the committee by the Ford Motor Company.[13] However, it is shorter, for example, than the list contained in the 1991 SRI study (Tables 11 and 12), and it excludes certain technologies (such as "friction reduction II") included in some studies by EEA (1990).[14]

Technologies are grouped into those that pertain to (1) the engine generally (internal friction, accessory loads, thermodynamic efficiency), (2) fuel systems, (3) valve train, (4) number of cylinders, (5) transmission technologies, and (6) technologies that affect power demands, such as rolling resistance, aerodynamics, and vehicle weight. These groups of technologies are mutually exclusive (e.g., 5-speed manual and 4-speed automatic transmissions) or mutually compatible (e.g., improved aerodynamics and advanced tires). Within mutually exclusive groups, market shares must sum to 100 percent; within mutually compatible groups, market shares are independent and each could be as high as 100 percent.

The EEA and SRI studies deal with passenger cars and generally exclude light trucks (EEA, 1988). However, nearly all of the technologies applicable to automobiles are applicable to light trucks, with a few exceptions and adjustments. For example, front-wheel drive is not likely to be as widely applicable to light trucks, in part because rear-wheel drive allows greater towing capacity. On the other hand, diesel engines are not included in the passenger-car options, but they are included for light trucks, for which problems of consumer acceptance and emissions standards are relatively less severe in certain market segments.

The costs of fuel economy improvement proved to be the most controversial factor in the analysis. Disagreements over costs have partly to do with valuing sunk investments. For example, one should compare the costs of *two* alternative *new* technologies by taking into account total capital costs, variable material and labor costs, and all overhead, burdens, and profit margins for each. The cost difference between the two would yield the expected difference in retail price equivalent (RPE).[15] On the other hand, in considering the cost implications of *replacing* a production facility for an existing technology with one for a new technology, one would compare the full capital

---

[13]Letter from Allan D. Gilmour, president, Ford Automotive Group, to Richard A. Meserve, chairman, Committee on Fuel Economy of Automobiles and Light Trucks, August 14, 1991.

[14]Although the types of technologies envisioned by EEA as "friction reduction II" are very likely to be adapted, they do not strictly satisfy the conditions for proven technologies. In the committee's view, these are the kind of probable technology improvements that might make the shopping cart assessment conservative, despite the relatively high maximum-penetration rates used for the proven technologies.

[15]The RPE is an estimate of the incremental retail cost to the consumer of purchasing a vehicle incorporating the new technology.

costs of the facility for the new technology with the lower level of capital spending required to maintain the manufacturing capability for the existing technology.[16] In such a case, the new technology would look far more expensive than the continuation of the existing technology. This difference is important because it relates to the timing of regulatory standards. To the extent that standards require the premature replacement of useful capital, their cost will be higher than if they allowed new fuel economy technology to be introduced as part of a "normal" capital turnover process.

The committee believes that part of the differences between the cost estimates provided by SRI and EEA can be explained by the timing of the required investments. SRI developed two sets of cost estimates, one for MY 1995 and another for MY 2001. (The estimates for MY 2001 are used in this study.) The estimates purportedly reflect the incremental costs that domestic manufacturers would face in making technology changes by 2001, given their current product plans. That is, the capital cost of a technology a manufacturer already has in production will be relatively lower than that of a new technology, for which production facilities and equipment may have to be purchased. On the other hand, the EEA analysis assumes that full capital costs must be borne for the baseline as well as the new technology, so that the difference in costs is likely to be smaller. The EEA estimates might be construed as reflecting the long-run RPE of a fuel economy technology in comparison to the RPE of a "base" technology. In contrast, SRI's estimates might be seen as short-run RPEs, in that they reflect the costs of premature abandonment of productive capital investments.

Results

Using the shopping cart approach, the committee developed two projections of fuel economy and costs for each vehicle size class. The projections rely on the committee's adaptation of SRI's and EEA's technology-specific estimates of fuel economy improvement potentials and costs and on the committee's assumptions regarding maximum market shares. The committee's projections are called Case A and Case B, respectively. Cases A and B are intended to indicate a range of possibilities, from relatively high-cost, moderate fuel economy improvements (Case A, based heavily on SRI data) to relatively low-cost, greater fuel economy gains (Case B, based heavily on EEA data). The resulting projections differ significantly from earlier ones by SRI and EEA because of the committee's market share assumptions and its adjustment of the SRI and EEA data in certain cases.

Figures 7-4 through 7-11 illustrate the results of the shopping cart analysis for the various passenger-car and light-truck classes.[17] The curves for Case A and Case B for

---

[16]The greater capital costs might be offset by higher labor productivity or lower materials costs in the new facility, but probably not by nearly enough to change the overall conclusion.

[17]Also shown on each figure for convenience are the trend and BIC projections.

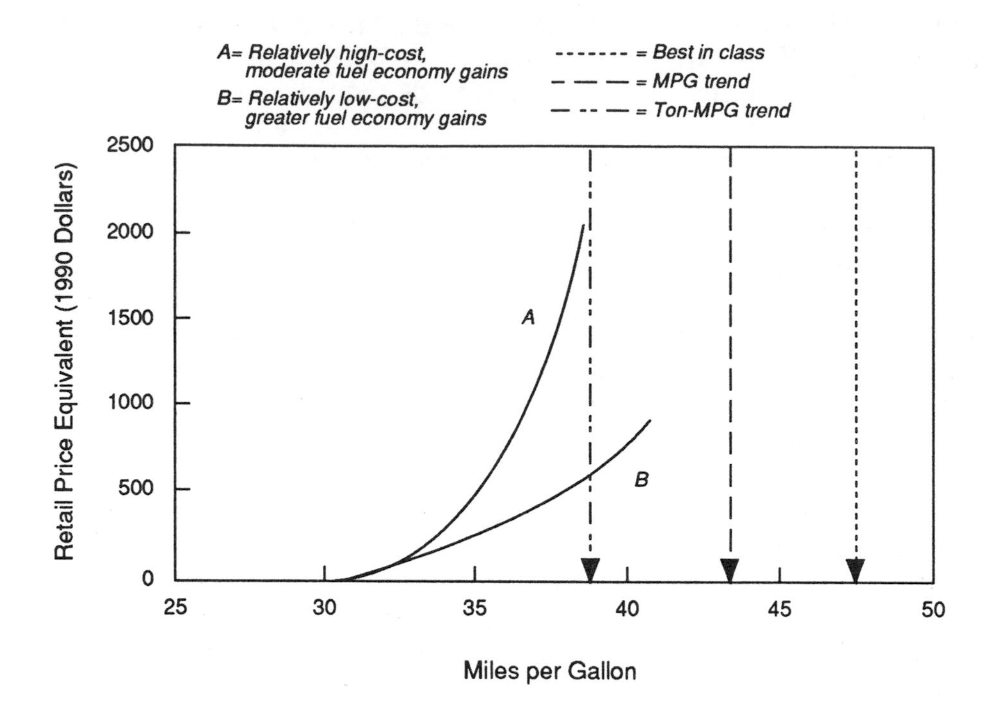

FIGURE 7-4 Shopping cart projection results for subcompact passenger cars.

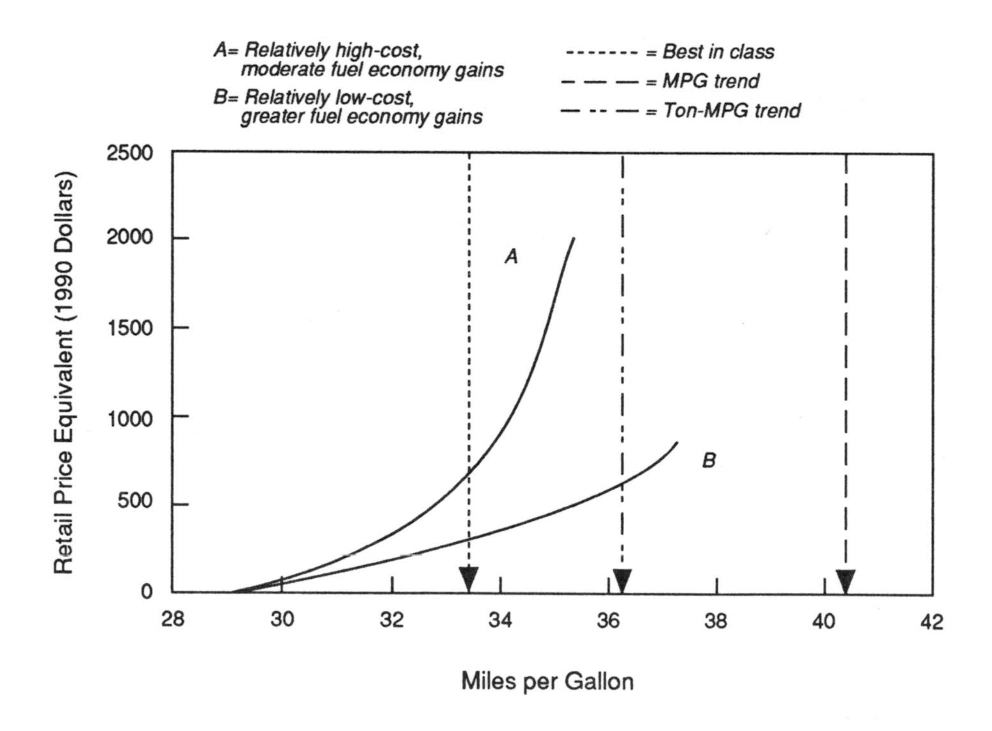

FIGURE 7-5 Shopping cart projection results for compact passenger cars.

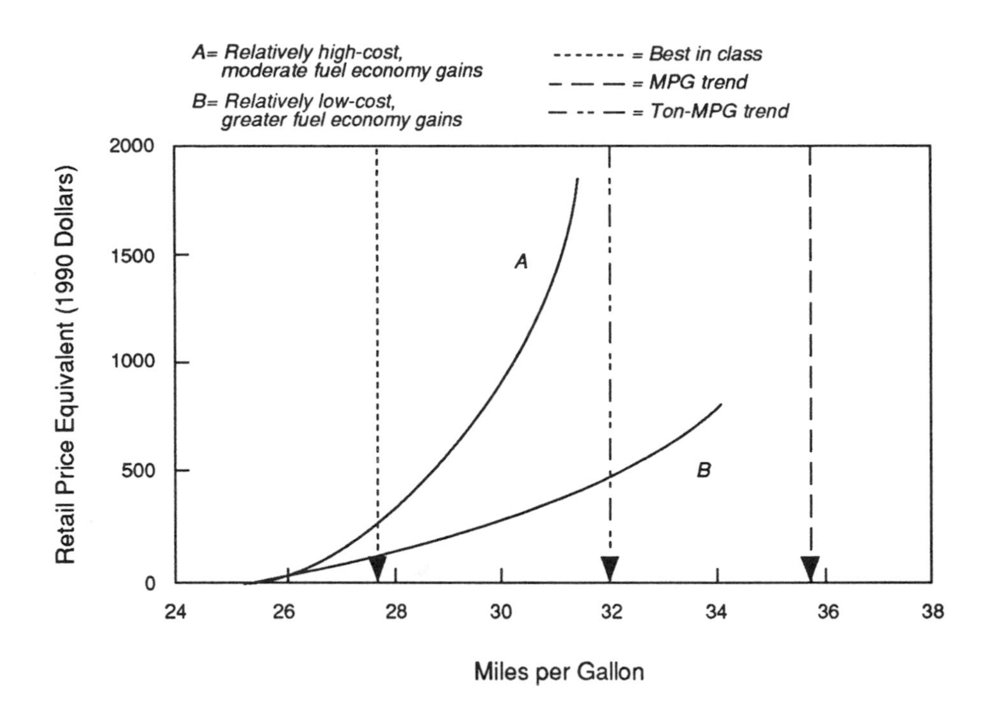

FIGURE 7-6 Shopping cart projection results for midsize passenger cars.

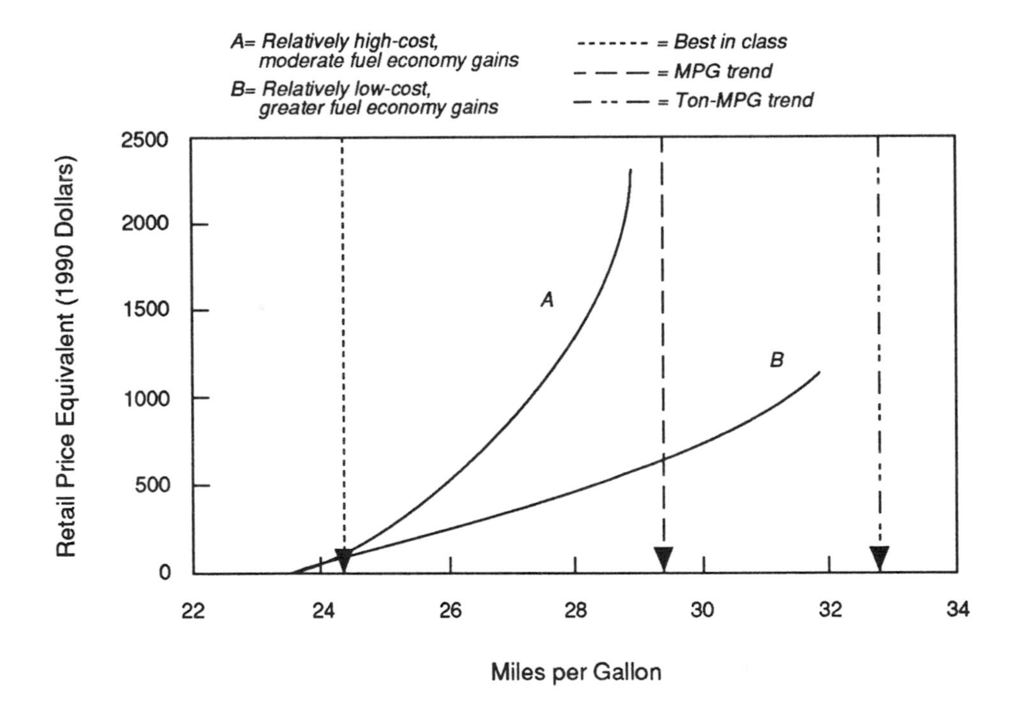

FIGURE 7-7 Shopping cart projection results for large passenger cars.

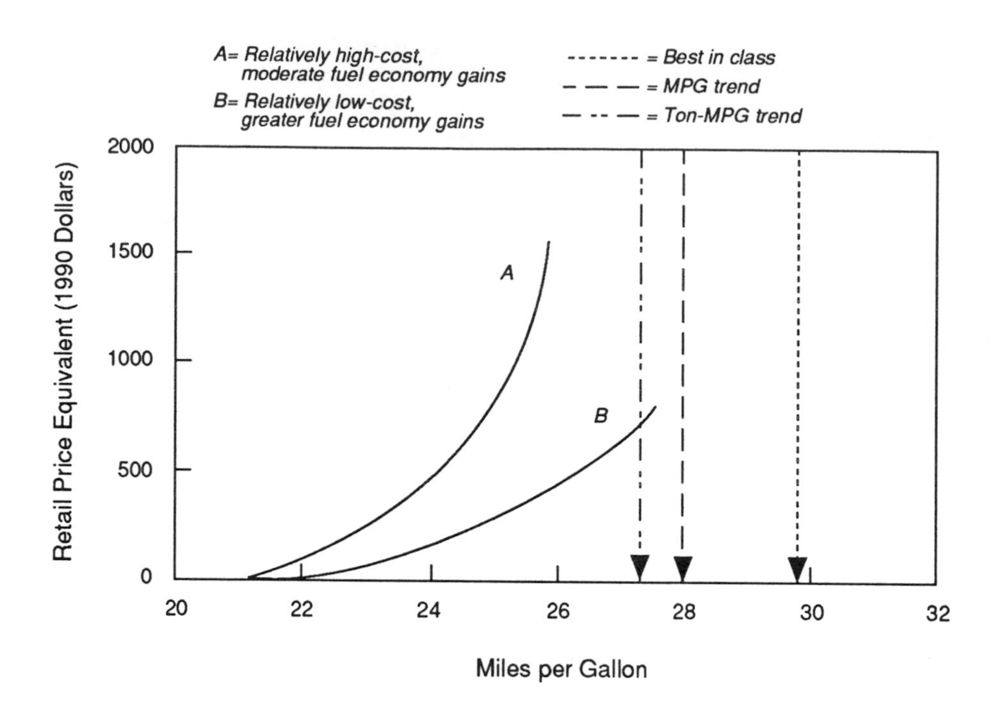

FIGURE 7-8 Shopping cart projection results for small utility vehicles.

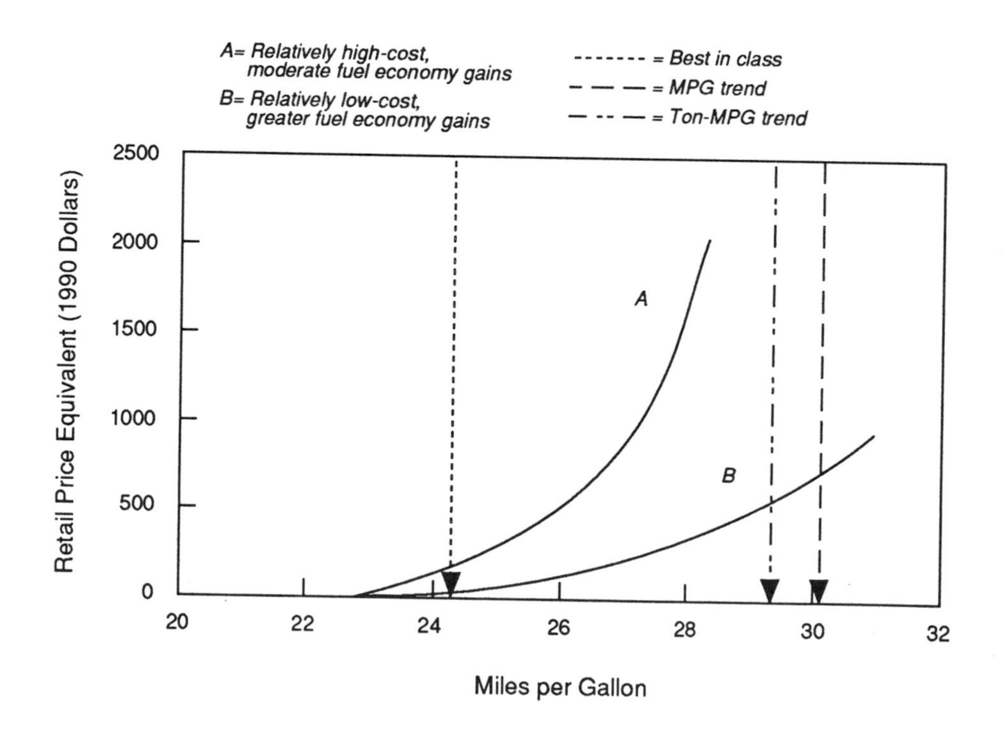

FIGURE 7-9 Shopping cart projection results for small vans.

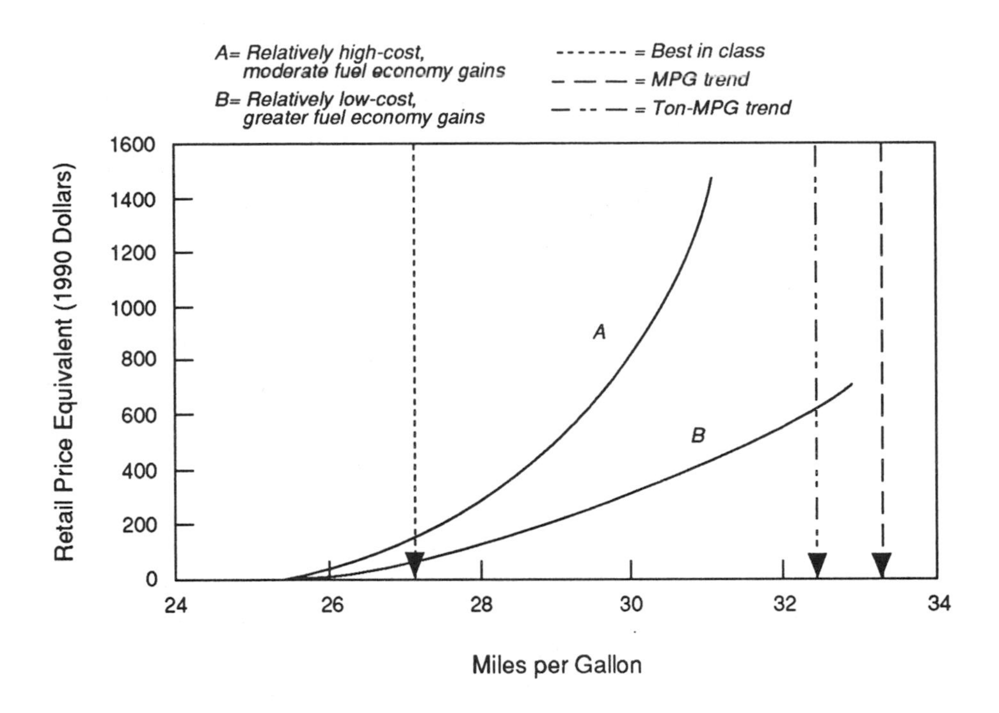

FIGURE 7-10 Shopping cart projection results for small pickups.

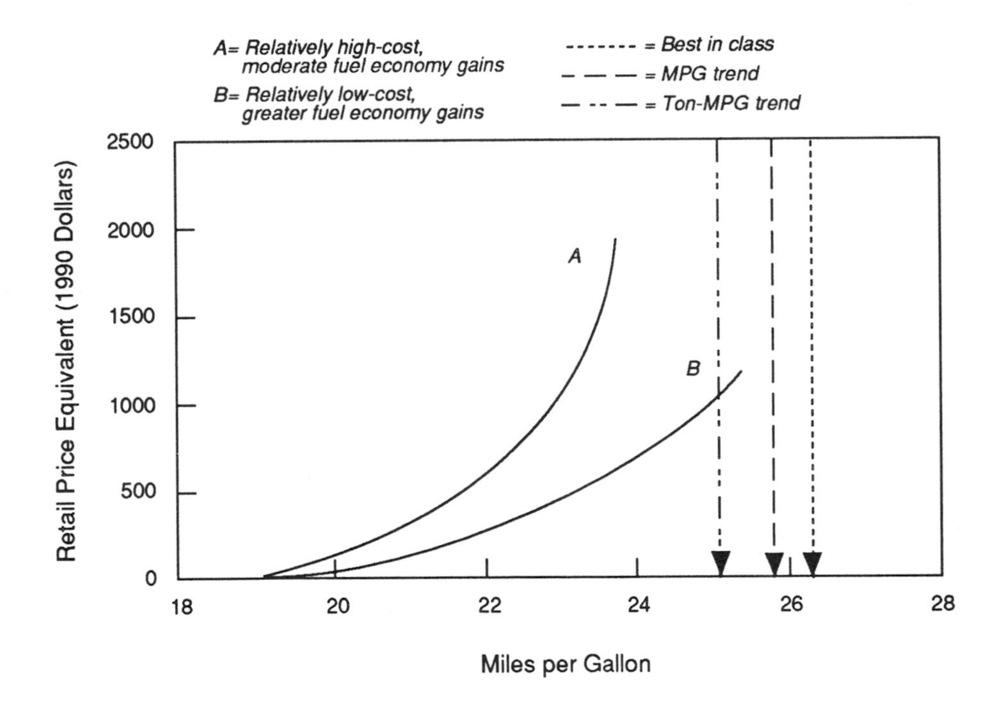

FIGURE 7-11 Shopping cart projection results for large pickups.

each class exhibit the expected concave upward shapes that result from adding technologies to the MY 1990 baseline in the order of their decreasing cost-effectiveness. For each vehicle class, the costs at a given level of fuel economy differ quite substantially for the two cases, but the end points (the levels of fuel economy at which the potential contributions of all proven technologies on the list are exhausted) differ much less. Put another way, the committee's shopping cart analysis yields cost projections that are inherently much more uncertain than the fuel economy projections. Finally, it should be noted that time is not a variable in Figures 7-4 through 7-11 and that the analysis is not intended to suggest that the time path of adoption of new fuel economy technologies would follow the curves of Case A or Case B. What the curves do represent is the locus of "least-cost" combinations of technologies that would give a class-average fuel economy of a particular level, subject to the many assumptions of the method.

To estimate the potential impact of performance reductions on the end points of the shopping cart curves, the committee increased the end-point fuel economy for each vehicle class by a percentage reflecting the changes associated with returning the MY 1990 horsepower/weight ratio for each class to its MY 1987 value. The results, summarized in Table 7-6, suggest that such performance reductions could facilitate an increase in end-point fuel economy ranging from 1 to 2 mpg for cars, and from 0 to 1.6 mpg for light trucks. The effects of performance reduction for each class are essentially the same whether based on Case A or Case B.

As seen by examining Figures 7-4 through 7-11, the various projection methods do not provide entirely consistent results. Sometimes the trend projection gives the highest fuel economy estimate, and sometimes the best-in-class approach suggests the highest level. While there is some tendency for the various projections to cluster, the methods do not provide clear and uniform estimates of future levels of automobile and light-truck fuel economy. The results are revealing, but they do not eliminate the need for judgment.

CONCLUSIONS

- Three very different methods were used to project the future fuel economy levels for passenger cars and light trucks. Each method exemplifies a different view of the problem and each yields different projections of the potential for fuel economy improvement by vehicle class. *None of the projections in this chapter reflects the views of the committee regarding practically achievable levels of fuel economy, which are discussed in the next chapter.*

- None of the projection methods predicts what *will* happen or what *ought* to happen. Each gives a different view of what could be achieved and provides a different insight about the implications of increased fuel economy.

TABLE 7-6 Effect of Performance Reduction on Projected Fuel Economy Levels

Vehicle Class	Performance Reduction[a] (percent)	Projected End-Point Fuel Economy[b] (mpg)		Increase in Fuel Economy Due to Performance Reduction[c] (mpg)	
		Case A	Case B	Case A	Case B
Cars					
Subcompact	12.7	38.5	40.7	1.8	1.9
Compact	13.8	35.3	37.2	1.8	1.9
Midsize	11.9	31.4	34.1	1.4	1.5
Large	8.6	29.0	31.9	0.9	1.0
Light trucks					
Small pickups	13.5	31.1	32.9	1.5	1.6
Small vans	0.0	28.3	30.9	0.0	0.0
Large pickups	12.7	23.8	25.4	1.1	1.2
Small utility vehicles	14.3	25.9	27.5	1.3	1.4

[a] Performance reduction is measured by changes in class-average horsepower/weight ratios from MY 1990 to MY 1987 levels from Heavenrich et al. (1991).

[b] Fuel economy projections are based on the shopping cart approach keeping vehicle performance fixed at MY 1990 levels.

[c] Each 1 percent decrease in the horsepower/weight ratio is estimated to result in a 0.38 percent increase in fuel economy, based on estimates of the effect of engine displacement on fuel economy (EEA, 1991a).

- Each method provides an estimate of what might be achieved 10 to 15 years into the future under normal rates of product redesign, capital investment, and vehicle sales. All three methods indicate that substantial fuel economy improvements may be technically possible.

- The technology-penetration method suggests that further accelerating the process of fuel economy increase could be quite expensive.

- The BIC-based projections of the percentage improvements in fuel economy vary greatly from class to class, which suggests that the method may not be reliable at the class level.

None of the projections offers proof of technical potential. Together, however, they increase confidence that substantial fuel economy improvements could be achieved over the next 15 years without compromising the functionality of light-duty vehicles. There will be a price to pay, however, and some trade-offs will be required. Whether attaining such levels of fuel economy is likely to be worth the cost to consumers, manufacturers, or the nation is the subject of the following chapter.

REFERENCES

Berger, J.O., M.H. Smith, and R.W. Andrews. 1990. A system for estimating fuel economy potential due to technology improvements. Paper presented at workshop of the Committee on Fuel Economy of Automobiles and Light Trucks, Irvine, Calif., July 8-12. University of Michigan, Ann Arbor.

Bussmann, W.V. 1990. Potential Gains in Fuel Economy: A Statistical Analysis of Technologies Embodied in Model Year 1988 and 1989 Cars. Chrysler Corporation, Detroit, Mich.

Chandler, W.U., H.S. Geller, and M. Ledbetter. 1988. *Energy Efficiency: A New Agenda*. Washington, D.C.: American Council for an Energy-Efficient Economy.

Energy and Environmental Analysis (EEA), Inc. 1988. *Light Duty Truck Fuel Economy: Review and Projections*, 1980-1995. Prepared for U.S. Department of Energy, Office of Policy, Planning and Analysis. Arlington, Va.

Energy and Environmental Analysis (EEA), Inc. 1990. Analysis of the Fuel Economy Boundary for 2010 and Comparison of Prototypes. Draft final report. Prepared for Martin Marietta, Energy Systems, Oak Ridge, Tenn. Arlington, Va.

Energy and Environmental Analysis (EEA), Inc. 1991a. Fuel economy technology benefits. Presented to the Technology Subgroup, Committee on Fuel Economy of Automobiles and Light Trucks, Detroit, Mich., July 31.

Energy and Environment Analysis (EEA), Inc. 1991b. Documentation of Attributes of Technologies to Improve Automotive Fuel Economy. Prepared for Martin Marietta, Energy Systems, Oak Ridge, Tenn. Arlington, Va.

Heavenrich, R.M., J.D. Murrell, and K.H. Hellman. 1991. *Light-duty Automotive Technology and Fuel Economy Trends Through 1991*. Control Technology and Applications Branch, EPA/AA/CTAB 91-02. Ann Arbor, Mich.: U.S. Environmental Protection Agency.

Ledbetter, M., and M. Ross. 1990. Supply curves of conserved energy for automobiles. *Proceedings of the 25th Intersociety Energy Conservation Engineering Conference*. New York: American Institute of Chemical Engineers.

Office of Technology Assessment (OTA), U.S. Congress. 1991. *Improving Automobile Fuel Economy: New Standards, New Approaches*. Washington, D.C.: U.S. Government Printing Office.

Samples, D.K., and R.C. Wiquist. 1978. *TFC/IW*. Presented at the International Fuels and Lubricants Meeting, Royal York, Toronto, November 13-16. SAE Technical Paper Series 780937. Warrendale, Pa.: Society of Automotive Engineers.

SRI International. 1991. *Potential for Improved Fuel Economy in Passenger Cars and Light Trucks*. Prepared for Motor Vehicle Manufacturers Association. Menlo Park, Calif.

U.S. Department of Transportation. 1991. *Briefing Book on the United States Motor Vehicle Industry and Market, Version 1*. Cambridge, Mass.: John A. Volpe National Transportation Systems Center.

Williams, L.S., and P.S. Hu. 1991. *Highway Vehicle MPG and Market Shares Report*. ORNL-6672. Tenn.: Oak Ridge National Laboratory.

8

ACHIEVABLE FUEL ECONOMY LEVELS

The charge to the committee was to estimate "practically achievable" fuel economy levels for new cars and light trucks by size class. Determining practically achievable fuel economy levels, however, necessarily involves balancing an array of societal benefits and costs, while keeping in mind where the costs and benefits fall. Such judgments must include a complex manifold of considerations, such as the financial costs to consumers and manufacturers, the impact on employment and competitiveness, the trade-offs of fuel economy with occupant safety and environmental goals, and the benefits to our national and economic security of reduced dependence on petroleum. In the committee's view, while this balancing should be supported by technical analysis, its execution is properly the domain of the political process. This committee cannot define the appropriate balance.

The committee has thus approached its charge by first seeking to estimate future "technically achievable" fuel economy levels by size class. As explained below, these estimates provide guidance, subject to certain assumptions, on the fuel economy that could be achieved using current technology. The committee also sets out its judgment of the range of retail price increases for new vehicles attributable strictly to such fuel economy improvements. (These estimates do not include price increases to meet occupant safety and emissions-control requirements.) **The technically achievable fuel economy levels should not be taken as the committee's recommendations on future fuel economy standards.** Rather, they should be viewed by policymakers as one of the many inputs necessary for determining what would be practically achievable in the coming decade.

The committee believes that the practically achievable levels--the levels of fuel economy for each size class that achieve an appropriate balance of a broad array of costs and benefits--are likely to be found in the regions between the levels that would be achieved without any governmental intervention and the "technically achievable" levels. It remains for policymakers to determine the form of any future regulations and the levels of fuel economy--the "practically achievable" levels--that in their judgment provide the appropriate balance of costs and benefits to consumers, manufacturers, and the nation as a whole.

This chapter opens with a discussion of technically achievable fuel economy. It then discusses some of the costs and benefits that might follow if the technically achievable levels were reached. Where feasible, these costs and benefits are quantified, but much of the discussion is qualitative. Also, where possible, the chapter illustrates the uncertainties in the fuel economy projections and in their impacts.

TECHNICALLY ACHIEVABLE FUEL ECONOMY

Method and Assumptions

The committee's determination of "technically achievable" levels of future fuel economy was accomplished using a structured judgmental process. The estimates are based in part on the projections reported in Chapter 7. They represent the collective professional judgment of the committee in light of the available evidence, taking into account various opportunities and constraints, as follows.

First, the committee assumed that all vehicles in the future fleet will have to comply with known safety standards and Tier I emissions standards under the Clean Air Act amendments of 1990. The expected impact of such standards on fuel economy was taken into account. (Tier II and California's standards were not explicitly considered; see Chapter 4.)

Second, the committee assumed that other vehicle characteristics of importance to consumers, such as interior volume, acceleration performance, and amenities will be essentially equivalent to model year (MY) 1990 vehicles. (The effects of performance reduction were considered separately.)

Third, to provide a measure of assurance that the levels could in fact be achieved, the committee assumed that technology that is currently used in mass-produced vehicles somewhere in the world will be available. In this regard, the committee members were not necessarily constrained in their individual judgments to consider only the list of "proven" technologies discussed in Chapter 2. Some members considered technologies now used largely to achieve performance enhancement that might otherwise be redesigned to achieve higher fuel economy; others considered technologies that offer a combination of relatively high-cost fuel economy improvement and other valued attributes. The range of technologies considered by the committee is such that they would pay for themselves at gasoline prices of $5 to $10 per gallon or less.

Aside from the limits imposed by these assumptions, no consideration of affordability, sales, employment, competitiveness, or safety impacts entered into the determination of the "technically achievable" fuel economy levels. All such matters should be considered, however, as policymakers establish the *practically* achievable levels.

Results

Based on its collective knowledge and judgment, and drawing on the results of the projections of future fuel economy displayed in Figures 7-4 through 7-11, the committee estimated the "technically achievable" fuel economy levels in MY 2006 for eight classes of passenger cars and light trucks. The results are shown in Table 8-1 as ranges over which the committee has varying degrees of confidence. The committee has a relatively high degree of confidence that the lower estimates of technically achievable fuel economy could be reached, given the underlying assumptions. The committee has less confidence that the higher fuel economy estimates in the various size classes could be achieved, and it believes that it would be risky to set fuel economy targets at or above these levels. The committee sought to narrow the ranges as much as possible, but enough uncertainty remains that it would be misleading to collapse the ranges to single point estimates.

Table 8-1 also presents the committee's judgments about the likely increases in cost (in terms of their retail price equivalents, or RPE) associated with each estimate of technically available fuel economy for each vehicle class for MY 2006.[1] Expected increases in the average prices of new cars and light trucks in MY 2006 associated with the technically achievable levels range from a low of $500 to a high of $2,750 in 1990 dollars. The wide range associated with each fuel economy level reflects the high degree of uncertainty the committee had to contend with in considering the costs of adopting fuel-saving technologies.

Table 8-2 presents the committee's estimates of technically achievable fuel economy levels for MY 1996. These estimates reflect the committee's view that, because of the long manufacturing lead times required and the sunk investments in existing facilities, little can be accomplished by MY 1996 beyond that now expected from the manufacturers' product plans without generating excessive cost and disruption in the automotive industry.[2] As discussed in Chapter 5, there is no realistic opportunity between now and 1996 to introduce new technology or increase the market penetration of existing technology in a fashion that will significantly increase fuel economy without extraordinary cost. More can be accomplished by MY 2001 and MY 2006 because new products developed for those years can more effectively reflect an increased emphasis on fuel economy.

---

[1]These cost/price estimates take into account not only the costs associated with the specific shopping cart projections in Chapter 7, but also the committee members' independent understanding of the costs of reaching the technically achievable fuel economy levels. The various cost estimates are an average for all vehicles in each class--any specific vehicle model might be subject to lower or higher cost increases than these. Moreover, for purposes of this analysis, the committee attributes all the costs of a technology to fuel economy. Some fuel economy technologies, however, may offer other benefits to the consumer. If so, the analysis exaggerates the cost of higher fuel economy.

[2]The MY 1996 projections reflect a gain of 0.7 miles per gallon (mpg) in fleet-average car and light-truck fuel economy over 1991 levels based on an estimate by SRI (1991:2) of 0.7 mpg for MY 1995 over MY 1990.

TABLE 8-1 "Technically Achievable" Fuel Economy for MY 2006 Vehicles

Vehicle Size Class	Ranges of "Technically Achievable" Fuel Economy in MY 2006[a/] (mpg)		Incremental Retail Price Equivalent for Improved Fuel Economy in MY 2006[b/] (1990 Dollars)	
	Higher Confidence	Lower Confidence	At Higher Confidence Fuel Economy	At Lower Confidence Fuel Economy
Passenger Cars				
Subcompact	39	44	500-1,250	1,000-2,500
Compact	34	38	500-1,250	1,000-2,500
Midsize	32	35	500-1,250	1,000-2,500
Large	30	33	500-1,250	1,000-2,500
Light Trucks				
Small pickup	29	32	500-1,000	1,000-2,000
Small van	28	30	500-1,250	1,000-2,500
Small utility	26	29	500-1,250	1,250-2,500
Large pickup	23	25	750-1,750	1,500-2,750

[a/] The term "technically achievable" is circumscribed by the following assumptions made by the committee. The estimates result from consideration of technologies currently used in vehicles mass produced somewhere in the world and that pay for themselves at gasoline prices of $5 to $10 per gallon or less (1990 dollars). The estimates assume compliance with applicable known safety standards and Tier I emissions requirements of the Clean Air Act amendments of 1990. Compliance with Tier II and California's emissions standards has not been taken into account. The estimates also assume that MY 2006 vehicles will have the acceleration performance of, and meet customer requirements for functionality equivalent to, 1990 models. The estimates take into account past trends in vehicle fuel economy improvements and evidence from "best-in-class" fuel economy experience. The term "technically achievable" should not be taken to mean the technological limit of what is possible with the current state of the art; nor should the committee's estimates of what is technically achievable be taken as its recommendations as to what future fuel economy levels should be.

Aside from the limits imposed by the foregoing assumptions, no cost-benefit considerations entered into the determination of the technically achievable fuel economy levels. Specifically, the estimates do not take into account other factors that should be considered by policymakers in determining any future fuel economy regulations, including impacts on the competitiveness of automotive and related industries, sales and employment effects, petroleum import dependence, effects on nonregulated emissions (e.g., the greenhouse gas, carbon dioxide), and the development and adoption of unanticipated technology.

As a point of reference, the Environmental Protection Agency's (EPA's) composite average fuel economy for MY 1990 passenger cars and light trucks, by size class, was as follows: passenger cars--subcompact, 31.4 mpg; compact, 29.4; midsize, 26.1; large, 23.5; light trucks--small pickup, 25.7; small van, 22.8; small utility, 21.3; large pickup, 19.1 (Heavenrich et al., 1991).

[b/] The retail price equivalents are estimates only of the incremental first cost to consumers of improved fuel economy. They do not include incremental costs associated with mandated improvements to occupant safety, which, on average for new passenger cars and light trucks, are expected to be $300 and $500, respectively in 1990 dollars; nor do they include incremental costs of controls to comply with Tier I emissions requirements, which are expected to range from a few hundred dollars to $1,600 per vehicle.

TABLE 8-2 "Technically Achievable" Fuel Economy Levels for MY 1996, MY 2001, and MY 2006

Vehicle Size Class	Fuel Economy (mpg)				
		MY 2001		MY 2006	
	MY 1996	Higher Confidence	Lower Confidence	Higher Confidence	Lower Confidence
Passenger Cars					
Subcompact	32	36	38	39	44
Compact	30	32	34	34	38
Midsize	27	29	31	32	35
Large	24	27	29	30	33
New-Car Fleet	29	31	33	34	37
Light Trucks					
Small pickup	26	28	29	29	32
Small van	24	26	27	28	30
Small utility	22	24	25	26	29
Large pickup	20	22	23	23	25
New Light-Truck Fleet	22	24	25	26	28

NOTE: See notes to Table 8-1. All estimates are rounded to nearest whole mile per gallon.

The new-car and light-truck average fuel economy by size class for MY 1996 assumes 0.7 mpg improvement in each size class from its corresponding EPA composite average level in MY 1991. This assumption is similar to that made by SRI (1991) for MY 1995 vehicles starting from a MY 1990 base.

Fuel economy values by size class shown for MY 2001 were obtained by interpolation between values for MY 1996 and MY 2006.

The new-car and light-truck fleet average fuel economies are shown above for illustrative purposes and are calculated assuming a size-class mix similar to that for MY 1990 vehicles, as follows: passenger cars--subcompact, 23 percent; compact, 35 percent; midsize, 26 percent; large, 16 percent; light trucks--small pickup, 15 percent; small van, 29 percent; small utility, 16 percent; large pickup, 40 percent. Data are based on Heavenrich et al. (1991).

After arriving at its consensus estimates for MY 2006 and making straightforward extrapolations from MY 1991 for MY 1996, the committee derived estimates for technically achievable fuel economy in MY 2001 by linear interpolation between MY 1996 and MY 2006 (see Table 8-2). This approach implicitly assumes a constant rate of progress between MY 1996 and MY 2006. In view of the judgmental nature of the MY 1996 and MY 2006 estimates, a more sophisticated approach to estimating the intermediate year levels was not warranted.

Using the estimated size-class fuel economy levels, the hypothetical fleet-average fuel economy estimates shown in Table 8-2 were calculated as a weighted harmonic average assuming that the size mix for the eight vehicle classes considered in this study remains the same as the MY 1990 mix. The committee's fleet fuel economy estimates for MY 2001 range from 31 to 33 mpg for passenger cars and from 24 to 25 mpg for light trucks. The comparable MY 2006 fleet averages are 34 to 37 mpg for passenger cars and 26 to 28 mpg for light trucks.[3]

Additional fuel economy increases could be obtained if the constraints and assumptions embodied in the definition of the "technically achievable" levels were relaxed. For example, it is clearly possible to trade off vehicle performance (e.g., the ratio of horsepower to weight, or hp/wt) for fuel economy. As discussed in Chapter 7, from 1987 to 1991, average hp/wt ratios increased by about 13 percent for all passenger cars and by 10 percent for all light trucks (size-class changes differ greatly). Assuming that a fuel economy improvement of 0.38 percent can be achieved for each 1 percent reduction in the hp/wt ratio (Energy and Environmental Analysis, Inc., 1991), reducing performance to MY 1987 average levels would yield an average gain in fuel economy of about 5 percent for passenger cars and 3.5 percent for light trucks, but at some loss of consumer satisfaction. Similarly, other vehicle attributes valued by consumers could be compromised in return for additional gains in fuel economy, at a cost of consumer satisfaction that is difficult to quantify, but potentially significant.

PRACTICALLY ACHIEVABLE FUEL ECONOMY

Cost-Benefit Considerations

As discussed above, determining the practically achievable levels of the future fuel economy of automobiles and light trucks requires consideration of the nature and magnitude of the costs and benefits of higher fuel economy, not only to consumers but also to industry, workers, and the nation as a whole. Upon consideration of these costs and benefits, policymakers may conclude that it is desirable to push fuel economy to the technically achievable levels defined above, or even beyond, or they may decide to set lower levels after considering all of the costs that might flow from higher levels.

---

[3]The committee assumed an unchanged fleet mix in order to make a rough estimate of the fleet averages. However, more stringent fuel economy regulations would raise the price of vehicles and induce mix shifting. Further, there is no reason to expect consumer preferences to remain fixed over the period. Thus, the MY 2001 and 2006 new-vehicle fleets would probably have a different mix from the MY 1990 fleet.

To assist policymakers in making these trade-offs, the committee has examined some of the costs and benefits, focusing particularly on those that can be illuminated by analyses based on the data and analyses used for other purposes in this study. Several complications are worthy of note at the outset, however.

First, the costs and benefits of achieving higher fuel economy are borne by different individuals and groups. For example, consumers and manufacturers may pay a financial price for more efficient vehicles. Yet the nation--even the world--may enjoy the benefits of reduced energy use and emissions of carbon dioxide (CO_2), as well as the benefits of enhanced national security. Moreover, some automotive manufacturers may be better situated than others to produce vehicles that meet tougher fuel economy standards so there may be differential impacts among them.

Second, the magnitude and incidence of the costs and benefits of achieving higher fuel economy depend on the precise level and form of the standards adopted, as well as on how manufacturers and consumers respond to the new requirements in the context of their overall market relationships. The projections of future fuel economy in Chapter 7 and the committee's estimates of technically achievable fuel economy reported earlier in this chapter were carried out only within vehicle classes, without reference to whether new standards might be applied at the class level or on some other basis.[4] As noted previously, this study did not examine the degree to which more stringent fuel economy standards might lead to downsizing, downweighting, or shifts in the market mix of vehicles of different classes, yet these factors are critical in defining the costs and benefits. Thus, policymakers are urged not to view the analyses of specific costs and benefits in this chapter as definitive, but rather as suggestive of further analysis that might be appropriate. Moreover, the quantitative values of the various costs and benefits depend on the levels of fuel economy, so they may differ from the values estimated here for the technically achievable levels of fuel economy.

Third, higher fuel economy imposes costs and benefits not only directly, but indirectly, on the various affected parties. To illustrate, to the extent that more stringent standards lead to an overall reduction in automotive fuel use, and thus to a reduction in overall national petroleum consumption, they may lead to a reduction in oil imports. This could, in turn, lead to an improvement in the nation's balance of payments and a lessening of national concern for the security of Middle Eastern oil supplies. While the nation as a whole might view this as a desirable set of circumstances, some groups--such as oil importers and shippers--might view it less favorably.

Fourth, some of the costs and benefits of higher fuel economy are quite uncertain. In some cases, this uncertainty yields the possibility of major miscalculations and errors. The sensitivity of the costs and benefits to the form of possible new standards exacerbates these possibilities. For example, some forms of corporate average fuel

---

[4]Chapter 9 discusses a range of possible types of future fuel economy requirements and offers qualitative remarks on their possible implications for some of the costs and benefits examined here.

economy (CAFE) standards could compel some manufacturers to make radical changes in the design and characteristics of their product lines. Should they misjudge future consumer preferences and "miss the market," they could be seriously harmed financially. On the other hand, fuel economy requirements that are set too low could leave the nation and its automotive industry seriously disadvantaged if world opinion embraces the concept that carbon dioxide emissions must be cut sharply and swiftly to forestall excessive global warming.

Finally, as fuel economy requirements are pushed ever higher, some of the benefits to the nation of higher standards, for example, reduced fuel use, become proportionally smaller. (The phenomenon of diminishing returns sets in, as suggested by Figure 8-1.) At the same time, some of the costs of higher fuel economy, as exemplified by the estimates obtained using the shopping cart method (Figures 7-4 through 7-11), become proportionally higher. Taking into account these kinds of effects, policymakers should be wary of setting fuel economy requirements too high.

Costs and Benefits of Higher Fuel Economy to Consumers

The estimates of technically achievable future vehicle fuel economy are based on the assumption that attributes valued by consumers, such as interior volume, comfort, acceleration performance, safety, and load-carrying capacity are not degraded. However, it was assumed that vehicle prices would have to increase to cover the costs of adopting technologies that could improve fuel economy while keeping these attributes fixed. The committee's estimates of these price increases are reported in Table 8-1.

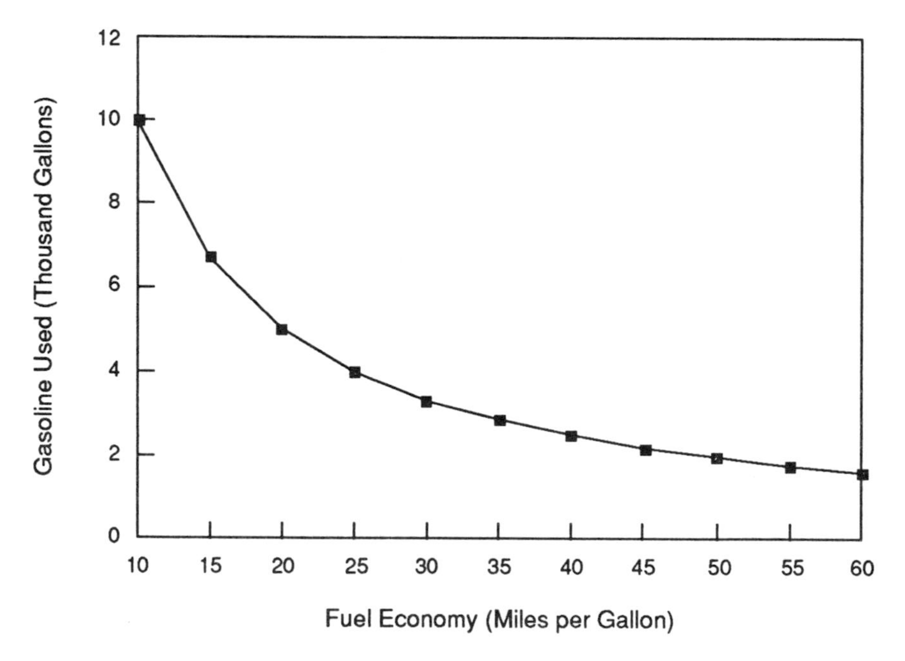

FIGURE 8-1 Dependence of fuel consumption on fuel economy for 100,000 vehicle miles traveled.

Higher fuel economy also benefits consumers by lowering the total amount they must pay for fuel during the lifetime of the vehicle. If the discounted present value of the vehicle lifetime savings in fuel costs exceeds the increase in the purchase price due to the incorporation of fuel economy technologies, the consumer will enjoy a net benefit. In theory, if all other attributes are the same, consumers should prefer a more expensive vehicle with higher fuel economy, so long as the discounted present value of the marginal fuel cost savings is greater than the increased initial purchase price.

Table 8-3 illustrates this principle using a simplified set of calculations of the lifetime fuel cost savings for each of the eight vehicle classes considered in this study. The vehicle price increases and class-average fuel economies in MY 2006 are those estimated by the committee as technically achievable.[5] The lifetime fuel cost savings are calculated under various assumptions of consumer discount rates and effective vehicle lifetimes. For these particular illustrations, it is assumed that the vehicle is driven 14,400 miles in its first year, declining by 5 percent per year thereafter; that the pump price of gasoline in the year 2006 is $1.45 per gallon in 1990 dollars, increasing 1.5 percent per year in real terms thereafter; and that consumers value future fuel cost savings over various time periods and discount them at various rates. No account is taken of differential resale value at the end of the vehicle lifetime,[6] or of the time value of money that consumers spend to make the initial purchase.[7]

Table 8-3 displays the results under various assumptions about the key parameters of the analysis. In a number of instances, the lifetime fuel cost savings exceed the initial purchase price increase, which means that higher fuel economy would offer a net benefit to the consumer under those conditions. Even when the net impact is a cost to the consumer, the net cost (price increase due to higher fuel economy minus fuel cost savings) is much lower than the increase in the vehicle price. It can also be seen from this table that the lifetime fuel cost savings are estimated to be about the same, whether the consumer is assumed to use a 30 percent discount rate and a 12-year decision horizon or a 10 percent discount rate and a 4-year decision horizon. By comparison, the fuel cost savings using a 10 percent discount rate and a 12-year decision horizon are nearly twice as high as those estimated under the other two sets

---

[5]The cost and benefits of practically achievable fuel economy levels would differ from those in Table 8-3.

[6]It is not clear whether the initial purchaser of a vehicle should take into account the fact that a vehicle with a higher fuel economy might have a higher value at time of resale or trade-in. Thus, it is also not clear whether the practical lifetime of a vehicle should be taken as the average time a typical vehicle is owned by its first owner or the average total lifetime of the vehicle.

[7]The latter assumption is equivalent to assuming that the downpayment plus the net present value of the time payments is equal to the purchase price.

TABLE 8-3 Effects of Technically Achievable Fuel Economy Improvements on Lifetime Fuel Cost Savings: Illustrative Example

Vehicle Class	Fuel Economy (mpg) 1990	Fuel Economy (mpg) 2006	Average Vehicle Price Increase (1990 $) High Estimate	Average Vehicle Price Increase (1990 $) Low Estimate	Average Lifetime Fuel Cost Savings [a] (1990 $) (vehicle life and discount rates) 12 Years, 30%	4 Years, 10%	12 Years, 10%
	Low Estimates of mpg in 2006						
Passenger Cars							
Subcompact	31.4	39	1,250	500	428	413	801
Compact	29.4	34	1,250	500	318	306	594
Midsize	26.1	32	1,250	500	467	470	912
Large	23.5	30	1,250	500	636	613	1,190
Light Trucks							
Small Pickup	25.7	29	1,000	500	305	295	571
Large Pickup	19.1	23	1,750	750	613	591	1,146
Small Van	22.8	28	1,250	500	562	542	1,051
Small Utility	21.3	26	1,250	500	586	565	1,095
	High Estimates of mpg in 2006						
Passenger Cars							
Subcompact	31.4	44	2,500	1,000	629	607	1,177
Compact	29.4	38	2,500	1,000	531	512	994
Midsize	26.1	35	2,500	1,000	672	646	1,257
Large	23.5	33	2,500	1,000	845	815	1,581
Light Trucks							
Small Pickup	25.7	32	2,000	1,000	529	510	989
Large Pickup	19.1	25	2,750	1,500	852	822	1,595
Small Van	22.8	30	2,500	1,000	726	700	1,359
Small Utility	21.3	29	2,500	1,250	860	829	1,609

[a] Present value of fuel cost savings discounted to 2006 for a new MY 2006 vehicle driven 14,400 miles in the first year, declining by 5 percent per year thereafter. Fuel price is $1.45 (1990 $) per gallon in 2006, increasing at 1.5 percent per year in real terms. Fuel savings are assumed to occur at the midpoint of each year and are based on the difference between the fuel economy of an average MY 2006 vehicle of each class and that of the average MY 1990 vehicle of the same class.

of assumptions.[8] Using either of the first two sets of assumptions, it is nearly always in the consumer's interest to pay at least the lower estimates of costs for these higher fuel economy levels. This probably overstates the value that consumers would actually give to energy-saving investments of this type, however.

The purchase price increase and fuel cost savings in each vehicle class depend on the level of fuel economy reached. As shown in Chapter 6, for reasonable estimates of the relevant parameters, the net consumer cost of high fuel economy is roughly constant over a range of several miles per gallon. This suggests that buyers of new vehicles will be indifferent to fuel economy at the present time, a stance seemingly supported by vehicle market trends over the past half decade.

Although consumers will receive some savings in fuel costs in return for the higher vehicle price associated with fuel economy improvements, it is uncertain whether the consumer will be satisfied with this trade-off. The net financial impact on the consumer depends on the prices of fuel and technology, both of which are very uncertain. Moreover, although the technologies considered by the committee do not require major changes in vehicle attributes, sometimes even subtle changes may make a large difference in consumer perceptions. Forced changes to meet stringent fuel economy standards increase the risk that a manufacturer may modify one or more product lines in ways that consumers will not find acceptable.

Safety and emissions regulations will also add to the price of new cars in the near future. Safety standards are expected to add $300 to passenger-car prices and over $500 to light-truck prices in coming years. Eighty percent of the cost is for front-seat airbags. Cost estimates for Tier I emissions controls vary widely, from a few hundred dollars to $1,600 dollars per car. Clearly, these cost estimates reflect a considerable range of uncertainty. The costs of Tier II standards are even less well understood, but are certain to be substantial (see Table 8-4). While these costs are independent of those arising from fuel economy improvements, the combined impacts of all regulations should be considered by policymakers when evaluating the benefits and costs of fuel economy, safety, and emissions regulations. Although consumers may perceive that the benefits they receive from safety improvements are commensurate with their costs, they may recognize that most of the benefits of pollution controls accrue to persons other than themselves, which may lead them to view their costs of meeting emissions reductions largely as a price increase.

Costs and Benefits of Higher Fuel Economy to Manufacturers

Improvements in the fuel economy of light-duty vehicles will affect vehicle sales and employment in the motor vehicle manufacturing industry. As noted above,

---

[8]There is some disagreement about the appropriate choices of these parameters. Research on consumer behavior tends to support the idea of a high discount rate and long decision horizon, whereas financial analysts tend to support the idea of a long decision horizon and lower discount rate, and some engineers favor a low discount rate and shorter time horizon. The first and third options always yield about the same results for net present value in these examples, and the second yields results nearly twice as great.

TABLE 8-4 Incremental Retail Prices for Improved Fuel Economy, Improved Occupant Safety, and Tier I Emissions Control in MY 2006 Vehicles (1990 dollars)

	Fuel Economy to Technically Achievable Levels[a]	Occupant Safety[b]	Tier I Emissions Control[c]
Passenger Cars	\$500 - \$2,500	\$300	From a few hundred dollars to \$1,600
Light Trucks	\$500 - \$2,750	\$500	

[a] The price ranges are over the four classes of passenger cars and four classes of light trucks asshown in Table 8-1.

[b] According to information provided to the committee by the National Highway Traffic Safety Administration, safety regulations will impose the following costs for passenger cars: \$257 for airbags, \$33 for side-impact improvements, and \$32 for additional head protection. Light-truck costs will be \$35 for rear lap belts, \$1 for steering changes, \$3 for headrest, \$1 for roof crush strength, \$11 for high-mount stop lamps, \$432 for airbags, \$26 for side-impact strength, and \$26 for head impact.

[c] Sierra Research (1988) estimated the cost of Tier I catalyst system improvements at \$139, and on-board vapor controls at \$25 per vehicle (other costs were not analyzed). In presentations to the committee, Ford reported estimates of \$275 per vehicle for on-board controls and \$550 for catalyst and any other tailpipe emissions improvements. Ford additionally estimated the cost of on-board diagnostic systems at \$300 and elimination of chlorofluorocarbons at \$125, for a total of \$1,250; General Motors estimated a total cost increase of \$1,600. EPA's estimates for Tier I standards are \$152 per car and \$57 per light truck; for on-board diagnostics, \$94 for cars and \$101 for trucks; for new evaporative emissions requirements, \$10 for cars, \$13 for light trucks. On-board vapor recovery during refueling is still under consideration. (Federal Register, 1990, 1991a, b.)

these effects would not fall uniformly on the industry. Depending on their form, new fuel economy regulations could affect some types of producers, such as full-line manufacturers heavily dependent on large vehicles, more than others.

Sales

Mandated higher fuel economy levels could have two major types of negative impacts on manufacturers. First, the higher cost of new vehicles is likely to depress sales if the resultant increase in the vehicle's initial price outweighs the benefit of fuel savings over the vehicle's life, as evaluated by consumers. Second, manufacturers may make design changes in attempting to raise fuel economy that turn out to be unsatisfactory to consumers. In the extreme, both of these effects would tend to reduce vehicle sales, reduce manufacturer profits, and reduce employment in the automotive industry and in sectors connected to the industry.

While it is nearly impossible to anticipate the design changes manufacturers might make and how consumers might respond to them, it is possible to make some estimates of the effect of increased vehicle costs on sales. Consider first the worst case--that consumers respond only to the increase in the retail price of the vehicle. Assume for sake of illustration that (1) a 1 percent price increase for a new car leads in the short run to a 1 percent drop in unit sales,[9] (2) that, in the absence of higher fuel economy standards, the average price of the MY 2006 car would be the same as the average price of a MY 1990 car, or $16,000 (MVMA, 1991); and (3) that the lower estimates of improved fuel economy are achieved. With these assumptions the retail price increases in Table 8-3 to this average vehicle price would yield unit sales reductions for new cars of about 3 to 8 percent, using the lower estimates of technically achievable levels. This would represent a significant impact on the automotive industry even though sales revenue would remain unaffected under the one-to-one elasticity assumption.

More realistically, however, consumers may take into account the benefits of reduced lifetime expenditures on fuel when making new-vehicle purchase decisions. In the absence of any estimate of the effect on vehicle sales of the present value of lifetime fuel savings (i.e., the "elasticity" of vehicle sales with respect to fuel costs), one way to make the estimate is to assume that vehicle sales respond to changes in the present value of fuel costs in the same way they do to vehicle price changes. As exemplified in Table 8-3, with the estimates of technically achievable fuel economy and costs reported for this study, the net present cost of higher fuel economy to the consumer is always substantially less than the initial vehicle price increase, and there may even be a positive benefit. Thus, the estimate of the impact on future vehicle sales given in the preceding paragraph is an upper bound that probably overstates the effect.

Employment

The impact of higher fuel economy on employment in automotive and related industries is difficult to assess. Under the worst case scenario for the impact on sales, discussed above, unit sales of passenger cars and light trucks might decline by 3 to 8 percent. Assuming that the number of jobs in the industry is proportional to unit sales (see Chapter 5), a reduction in sales of this magnitude would lead to a decline in employment of similar proportions.

Several factors would tend to mitigate this decline. First, as noted above, the worst case scenario assumes that consumers respond only to the higher price of more efficient vehicles. To the extent that this effect is offset by fuel cost savings, sales reductions--and thus employment reductions--would be less than the worst case suggests.

---

[9]An elasticity of -1 to -1.5 was reported by Michael J. Boskin in his presentation to the committee at the workshop that was held in Irvine, California, July 8-12, 1991. Ford Motor Company, in its presentation to the Impacts Subgroup of the committee, September 16, 1991, suggested an elasticity of -1.

Second, as also noted above, even a decline in unit sales due to a price increase might leave automotive industry sales revenues unchanged since higher prices might essentially offset lower unit sales. Some portion of the price increase, however, arises from the wages paid to the workers who manufacture and install the new fuel economy technologies that generate the price increase. So, while there might be a loss of jobs due to any shrinkage in unit sales, the loss would be offset to some extent by the growth of jobs in producing the fuel economy technologies.

The employment impacts of fuel economy changes must also be put in context. As pointed out in Chapter 5, the global automotive industry is now in a state of chronic overcapacity that is likely to continue for some time. Overcapacity, combined with long-term increasing productivity, will lead to continuing reductions in employment in motor vehicle manufacturing and related industries, regardless of fuel economy standards. Indeed, the long-term shrinkage in employment may dwarf the employment impacts of new fuel economy requirements.

Competitiveness

More stringent fuel economy standards may have a greater impact on who manufactures cars for the U.S. market and where they are manufactured than on aggregate sales or employment. That is, higher fuel economy standards may affect the competitive position of U.S. manufacturers vis-à-vis the rest of the world more than they would affect the industry as a whole. The importance of the competitive effect depends heavily on the form of the regulation, perhaps more so than its level. For example, a fuel economy standard that averages corporate fuel economy performance across all classes--the current approach--has important differential effects among manufacturers. Because domestic producers make a larger proportion of midsize and large cars than do most foreign manufacturers, even if domestic producers met the committee's technically achievable fuel economy levels in each class, they could not reach the same corporate average levels of fuel economy as the foreign (primarily Japanese) producers could unless they changed their market mix and incurred higher adjustment costs.[10]

Imported light trucks already achieve significantly higher fuel economy than domestic trucks in the same size classes.[11] If fuel economy standards for light trucks were set above levels characteristic of current domestic trucks, but at or below those of the imports, the domestic manufacturers would be at a serious disadvantage. In particular, import manufacturers could make their light trucks larger, heavier, or more powerful in each size class to gain market share while domestic manufacturers would be severely constrained.

---

[10]These matters are discussed more thoroughly in Chapter 9.

[11]For example, in 1990 the fuel economy of imported small pickups was 27.0 mpg, compared with 24.5 mpg for domestics, 34.1 vs. 21.2 mpg for compact utilities, 28.6 vs. 16.2 mpg for full-size utility vehicles, and 25.4 vs. 22.9 mpg for compact vans (Heavenrich et al., 1991).

Costs and Benefits of Higher Fuel Economy to the Nation

Petroleum Consumption

High levels of fuel economy would reduce the consumption of petroleum by the automotive fleet. For example, if today's new passenger-car fleet were to consume gasoline at the "technically achievable" average level of 34 to 37 mpg, the consumption of gasoline by this fleet would be reduced by about 18 to 26 percent, compared with the standard of 27.5 mpg.[12] Clearly reductions of this magnitude could enhance energy security, reduce CO_2 emissions, and provide some of the other benefits that are discussed in Chapter 1.[13] For example, fuel economy improvements of this magnitude, if attained today for the entire car and light-truck population, would reduce emissions of CO_2 from anthropogenic sources in the United States by an estimated 3 to 5 percent.[14] Reduced fuel use could lead to lessened U.S. dependence on imported oil, with attendant positive impacts on the balance of trade and on national security costs and risks.

Safety

It is the committee's view that the safety impact of fuel economy improvements of the levels projected in this report could be small. However, a number of factors prevent making this a definitive conclusion. Although fuel economy improvements could be achieved with little downsizing, it cannot be guaranteed that this approach will be followed. If consumers accept smaller cars, manufacturers may find downsizing a more profitable route to higher fuel economy than the introduction of costly technology that makes downsizing unnecessary. Thus, the committee concludes that there is likely to be some safety cost of higher fuel economy, and fuel economy measures that involve reductions in vehicle size or weight must be carefully monitored and analyzed to determine any safety consequences. But, as discussed more fully in Chapter 3, individuals and policymakers continually trade off safety and other values. The safety consequences should be one factor in the weighing of costs and benefits associated with achieving improved fuel economy of cars and light trucks.

---

[12]Of course, even if the fuel economy standards for new cars and light trucks were set at the "technically achievable" levels in 2006, such significant reductions would not be achieved until years thereafter. New vehicles constitute only a small portion of the existing fleet; the standards have impact only over time as the fleet turns over and more efficient new vehicles replace less efficient older vehicles. Further, increased fuel economy would probably increase vehicle miles traveled so the net gasoline savings would be somewhat less.

[13]Our failure to quantify all these benefits should not be construed to mean that the committee does not appreciate their significance. Rather, it reflects the difficulty in estimating and reliably evaluating them.

[14]Committee estimate based on data on sources of CO_2 from OTA (1991) and on the assumption that reaching the "technically achievable" level would result in an eventual 18 to 26 percent reduction in gasoline use for the entire automobile and light-truck fleet.

Emissions and the Environment

Fuel economy increases are likely to have little direct impact on vehicle emissions (except for CO_2), but emissions standards more stringent than the Tier I standards of the Clean Air Act amendments of 1990 are likely to have adverse effects on fuel economy. In addition, depending on the balance of price increase and fuel cost savings discussed earlier, increases in new-vehicle prices could keep older vehicles on the road longer, thus delaying reductions in total automotive emissions. For the estimates of "technically achievable" fuel economy presented in Table 8-1, it was assumed that safety and Tier I emissions regulations will be met, but Tier II and California's standards were not taken into account. If Tier II standards of California's standards are considered--especially the strict NO_x standards--it is likely that the application of several fuel-efficient engine designs (such as the lean-burn, the two-stroke, or the diesel engine) in light-duty vehicles could be precluded unless a major breakthrough occurs in emissions control for lean-burn engines.

The Risk of Choosing Incorrectly

As noted above, the benefits of fuel economy standards stem from the reduction in the consumption in petroleum. Figure 8-2 shows gasoline savings on a percentage basis as fuel economy is increased above 27.5 mpg. The noteworthy element of the curve is its non-linearity: as fuel economy is increased, less gasoline is conserved for each increase in mpg. Thus, improvement in fleet fuel economy offers diminishing returns on the benefit side of the equation.

The evaluation of costs to consumers is more complicated. As shown in Chapter 6, the curve of net costs to the consumer of improved fuel economy is relatively flat over a fairly broad range. (This flat portion is termed the "indifference region.") The gains or losses from improved fuel economy may amount to a few hundred dollars--a small amount in comparison to the $16,000 purchase price of the average new car (MVMA, 1991). Moreover, since evaluating the financial benefits of buying fuel economy technology requires anticipating uncertain future fuel prices and cost savings and discounting them to present value, it would not be surprising if consumers do not take fuel economy improvement into account. Because manufacturers may have to redesign their product lines to achieve improved fuel economy--a costly and risky proposition--they may decide to leave well enough alone and not risk making changes. Hence, in the absence of effective market pressures to encourage improved fuel economy, there are reasons to set fuel economy standards at the outer limits of the indifference region. Standards set at this boundary would not threaten extraordinary costs to consumers.

The difficulty in such an approach arises from the fact that the outer limit of the consumer indifference region is not easily determined. The combination of the rapidly rising costs of higher fuel economy levels and the diminishing returns to fuel savings per mile driven means that the net cost of higher fuel economy to the consumer increases rapidly outside the indifference region. Thus, the adverse consequences to consumers, manufacturers, and workers of overestimating the upper limit of the indifference region may be severe.

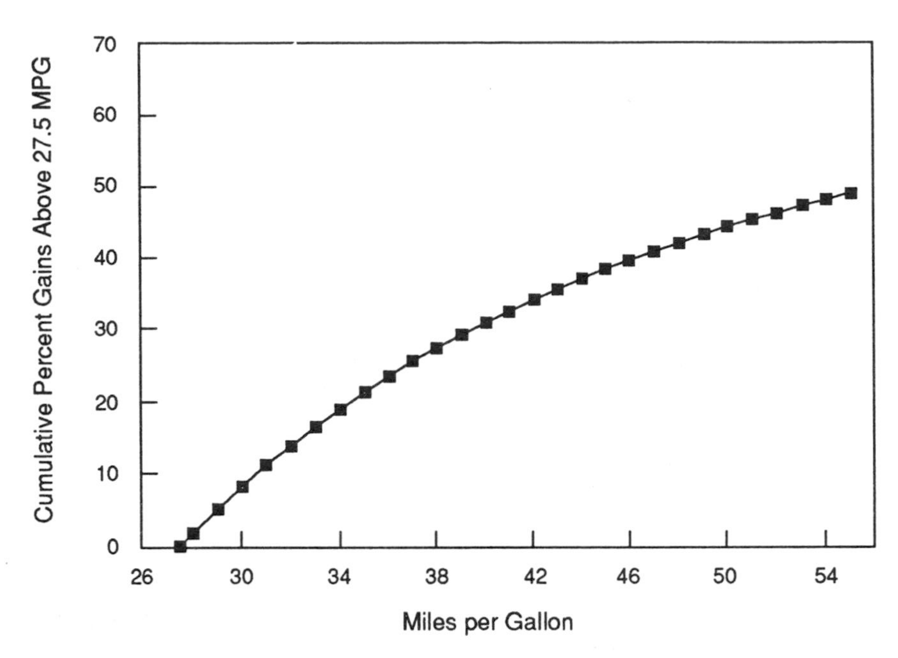

FIGURE 8-2 Cumulative percentage increase in fuel savings as a function of fuel economy with 27.5 mpg as a base.

Other risks are also associated with overly stringent fuel economy requirements. The more stringent the fuel economy requirement, the greater the likelihood that the manufacturers will be required to introduce significant modifications of existing cars in order to comply. If consumers do not like the design changes that are made to increase fuel economy, the cost in lost sales and profits to the manufacturers could be substantial, even if fuel savings outweigh the retail price increase to the consumer. And, stringent fuel economy requirements would impose special pressures on U.S. domestic manufacturers at a time when they are facing severe competitive and economic challenges from foreign competitors.

In sum, there are diminishing benefits and increased costs if fuel economy standards are set too high. This argues that policymakers should be wary of pushing fuel economy too far.

POLICY COORDINATION AND ANALYSIS

The interactions of higher fuel economy, improved occupant safety, and lower emissions illustrate dramatically the need for coordinated action by policymakers. There are limits to what can be achieved, and there are trade-offs that must be made. In setting fuel economy, safety, emissions, and other standards that affect automotive manufacturers, Congress and the administration should consider the cumulative impacts of the various regulations. Because the standards are set by different agencies under differing statutory commands, however, little coordination of policy is achieved and inconsistent pressures--such as those arising from environmental and fuel economy considerations--are allowed to persist. This is indeed unfortunate because the automotive industry accounts for significant employment and is the focus of much of the national debate on improvements to the economy, the environment, and competitiveness.

As noted above, the determination of the point at which the costs and benefits of improved fuel economy are in balance is surrounded by considerable uncertainty. Even the evaluation of the "technically achievable" fuel economy levels can only be roughly ascertained. And, the assessment of the appropriate balance is all the more uncertain because it requires trade-offs of incommensurate considerations: energy security, jobs, safety, environmental protection, and many other factors. There is a clear need for more information than the committee has had available to it.

If more analysis had been carried out by a greater variety of researchers, the uncertainty about the potential for fuel economy improvement and its costs would now be considerably lower. It is clear from the committee's analysis that the risks of setting overly stringent fuel economy standards could have annual costs to consumers and industry in the tens of billions of dollars, not to mention the possible costs in jobs and lives. On the other hand, insufficient fuel economy improvement could contribute to continuing energy insecurity and growing greenhouse gas emissions. The very useful contributions by analysts supported by the U.S. Department of Energy notwithstanding, federal support for data collection and analysis on this subject during the past decade has been very inadequate. In the committee's opinion, if the federal government intends to continue regulating automotive fuel economy, it must make a more substantial investment in understanding the technical issues, costs, benefits, and risks of such regulation.

REFERENCES

Energy and Environmental Analysis, Inc. 1991. Fuel economy technology benefits. Presented to the Technology Subgroup, Committee on Fuel Economy of Automobiles and Light Trucks, Detroit, Mich., July 31.

Federal Register. 1990. Vol. 55, No. 13 (Friday, January 19).

Federal Register. 1991a. Vol. 56, No.108 (Wednesday, June 5):25737.

Federal Register. 1991b. Vol. 56, No. 185 (Tuesday, September 24):48292-48293.

Heavenrich, R.M., J.D. Murrell, and K.H. Hellman. 1991. *Light-Duty Automotive Technology and Fuel Economy Trends Through 1991*. Control Technology and Applications Branch, EPA/AA/CTAB/91-02. Ann Arbor,Mich.: U.S. Environmental Protection Agency.

Motor Vehicle Manufacturers Association (MVMA). 1991. *Facts & Figures '91.* Detroit, Mich.

Office of Technology Assessment (OTA), U.S. Congress. 1991. *Changing by Degrees: Steps to Reduce Greenhouse Gases*. Washington, D.C.: U.S. Government Printing Office.

Salter, M.S., A.M. Webber, and D. Dyer. 1985. U.S. competitiveness in global industries: Lessons from the auto industry. In *U.S. Competitiveness in the World Economy*, B.R. Scott and G.C. Lodge, eds. Cambridge, Mass.: Harvard Business School Press.

Sierra Research, Inc. 1988. *The Feasibility and Costs of More Stringent Mobile Source Emission Controls*. Prepared for the Office of Technology Assessment, U.S. Congress. Sacramento, Calif.

SRI International. 1991. *Potential for Improved Fuel Economy in Passenger Cars and Light Trucks*. Prepared for Motor Vehicle Manufacturers Association. Menlo Park, Calif.

9

POLICIES FOR IMPROVING FUEL ECONOMY

Previous chapters have evaluated the opportunities for and constraints on the improvement of the fuel economy of automobiles and light trucks. This chapter confronts a different subject--the policies to require or induce lower fuel consumption. The existing corporate average fuel economy (CAFE) system is one such approach. The committee's investigations have revealed, however, that the existing system deserves careful reconsideration. Accordingly, this chapter discusses the CAFE system and some of its principal alternatives.

COMMENTS ON THE EXISTING CAFE SYSTEM

Congress has set in place a particular approach--the CAFE system--for improving the fuel economy of automobiles and light trucks (15 U.S.C. §§ 2001 et seq. (1988)). The statute set a fleet fuel economy standard for automobiles of 27.5 miles per gallon (mpg) for model year (MY) 1985 and thereafter, unless the limit was relaxed by the Secretary of Transportation (15 U.S.C. §§ 2002(a)(1), (4) (1988)).[1] Compliance is measured as the harmonic average of the fuel economy, as measured on a designated test cycle, of the automobiles produced by a manufacturer (15 U.S.C. § 2003 (1988)).[2]

---

[1]The standards for MY 1986-1988 were adjusted to 26 mpg, and for MY 1989 to 26.5 mpg. The standards for 1990 and thereafter have remained at the statutory target of 27.5 mpg (49 C.F.R. § 531.5 (1990)).

The legislation did not specify a particular standard for light trucks, but rather allowed the standard for light trucks to be determined by the Secretary of Transportation. 15 U.S.C. § 2002(b) (1988). The standards have varied over time; the standard for light trucks in 1991 was 20.2 mpg. 49 C.F.R. § 533.5 (1990).

[2]The harmonic average is the reciprocal of the average of the gallons per mile used by each car in the fleet. For example, a company that sells a vehicle with a fuel economy rating of 20 mpg and a vehicle with a fuel economy rating of 40 mpg would achieve a CAFE rating of 26.7 mpg. The use of the harmonic average is appropriate because it provides a measure of the fuel consumption of the fleet. That is, in the above example, if the two cars were each driven 10,000 miles, they would jointly consume 750 gallons of gasoline, thereby achieving an aggregate fuel economy of 26.7 mpg.

A manufacturer may have both a domestic and an import fleet and must separately comply with the limit for each.

The regulatory approach for improving fuel economy departs from the approach applied to automobile safety and environmental pollution. In these other areas, Congress has required that every vehicle satisfy certain minimal standards. The CAFE system, in contrast, allows vehicles with low fuel economy to be sold, so long as they are counterbalanced by the sale of vehicles with high fuel economy.[3] The system thus maintains a measure of flexibility for both manufacturers and consumers, which presumably reflects Congress's conclusion that, for example, the opportunity to purchase large cars should be preserved despite their low fuel economy. The net effect is that the CAFE system, while encouraging vehicle weight reduction and other changes, allows some diversity in the fleet.

As discussed earlier, the introduction of CAFE standards coincided with significant advances in automotive fuel economy. There is some disagreement among analysts as to the impact of CAFE standards in bringing about the change. Some of the increased fuel economy may be attributed, for example, to the impact of rising fuel prices in the late 1970s and early 1980s (see, e.g., Leone and Parkinson, 1990). Nonetheless, there is no doubt that the CAFE standards have, at the least, reinforced market pressures in bringing about an upgrading of the fuel efficiency of the vehicle fleet.

The committee finds, however, that the CAFE system has both weaknesses and strengths. Some of the weaknesses are outlined below:

- **Dissonance Between CAFE Requirements and Market Signals**. The CAFE approach places the manufacturer in an awkward position in the marketplace. The system constrains consumer choice--at least in the aggregate--to vehicles with characteristics that differ from those that otherwise would be offered. In effect, the manufacturers are required to sell vehicles with higher fuel economy regardless of consumer interest in purchasing such vehicles.[4] A rational consumer--other things being equal--should always prefer a fuel-efficient vehicle over a fuel inefficient vehicle. But, in fact, an increase in fuel economy can impose costs: the financial costs of the technology to achieve fuel economy and the costs reflected in reduced vehicle size, decreased performance, or other undesired attributes.[5] To the extent that improved fuel economy is achieved at the expense of other characteristics of the vehicle that the

---

[3]Because of the effect of harmonic averaging, the CAFE system can present the manufacturers with a more substantial challenge than might be immediately apparent. For example, to achieve a CAFE level of 30 mpg, a manufacturer would have to sell two 40-mpg vehicles for each 20-mpg vehicle that is sold.

[4]Surveys indicate that the typical consumer now values fuel economy far less than a variety of other vehicle attributes (presentation by Ford Motor Company to the committee's Impacts Subgroup, September 16, 1991; Consumer Attitude Research, 1991).

[5]As discussed in Chapter 6, reduced fuel consumption enables the consumer to recover some of the costs through lower operating costs. Moreover, some of the technologies that improve fuel economy provide other benefits to consumers. Nonetheless, the CAFE system is intended to require manufacturers to produce vehicles that depart from those that consumers might otherwise select.

consumer values more,[6] the manufacturer is being required to try to sell a product that does not reflect consumer demand. The consequence is an undesirable diminution in consumer satisfaction and, presumably, a reduction in overall automotive sales.[7] In short, over the past several years, there has been a growing dissonance between CAFE standards and market signals as the real price of gasoline has fallen.

- **Competitive Effects.** The existing CAFE standards can have perverse competitive effects. The standards have their most severe impact on the full-line manufacturers--the manufacturers that include large cars among their offerings. Because large cars in general have lower fuel economy than small cars, full-line manufacturers must invest resources to increase their sales of small cars and/or must invest in technology to increase the fuel economy of their large cars. Manufacturers of small cars, on the other hand, can more readily meet the standards and, indeed, may have fleets sufficiently above the CAFE standards as to enable them to expand initially into the large-car market without applying the expensive technology required of the full-line manufacturer. The system, thus, does not present equivalent technical or financial challenges to the manufacturers: Full-line manufacturers must strain to comply, whereas small-car manufacturers can comply with comparative ease. As it happens, this characteristic of the CAFE system has operated to the benefit of the Japanese manufacturers, and at the expense of the domestic (and some European) manufacturers. The CAFE system, thus, enhanced the competitive position of those foreign manufacturers that now pose the greatest threat to the domestic industry.

- **Attenuated Impact on Fuel Consumption.** Although the CAFE system encourages a fleet with higher fuel economy, it does not provide incentives for the *use* of a vehicle in a way that conserves fuel. In fact, to the extent that vehicles become more efficient and fuel prices remain unchanged, the consumer has incentives to respond to the lower per-mile fuel costs that efficient vehicles provide by driving more. Indeed, exactly such a trade-off has taken place: The improved fuel economy of automobiles has saved fuel, but some of the gains in fuel economy have been spent in the form of increased travel (cf. Chapter 6). To the extent that vehicle miles traveled (VMT) increase as a result of improved fuel economy, the goal of reduced overall fuel consumption is compromised.

---

[6] Some technologies that increase fuel economy can be attractive to consumers. For example, aerodynamic styling, four-valve-per-cylinder engines, and electronically controlled transmissions may enhance both fuel economy and consumer interest. To the extent that fuel economy requirements impose constraints on the manufacturers, however, the net effect is a reduction in the manufacturers' ability to optimize vehicles in a way that maximizes consumer satisfaction.

[7] The CAFE system initially imposed constraints on the automotive manufacturers at a time in the late 1970s and early 1980s in which the American public was experiencing a rapid increase in gasoline prices. It may plausibly be argued that the CAFE system, by requiring the manufacturers to produce more fuel-efficient vehicles, in fact prepared the manufacturers to meet the consumer demand that arose for those vehicles as a result of the increased fuel prices. The late 1980s, however, saw a reduction in gasoline prices in real terms. As a result, the achievement of high fuel economy is now of lesser importance to the public than other vehicle characteristics.

- **Slow Impact on Fuel Economy of Fleet.** The CAFE standards affect only a portion of the fleet--the new vehicles sold in the years to which the standards apply. Even aggressive CAFE requirements, thus, have a slow impact on overall fuel consumption because of the slow turnover of the vehicle fleet. Moreover, if the standards became so stringent that the cost of automobiles increases significantly or their characteristics (e.g., size, performance) became less desirable, consumers might tend to continue to operate older cars. Because the older vehicles in the fleet are in general the least fuel efficient and the most undesirable in terms of environmental pollution, incentives that inhibit vehicle turnover could delay the achievement of the objectives of the regulatory system.[8]

- **Gaming.** The CAFE regulatory approach--perhaps like any complex regulatory system--also invites "gaming" to exploit the system.[9] For example, because the fuel economy of automobiles is determined on a specified test cycle, manufacturers seeking to enhance their CAFE performance understandably adjust their cars--the shift points on automatic transmissions, for example--to optimize fuel economy on the test cycle. To the extent that the test cycle departs from actual driving conditions, however, cars tuned for fuel economy on the test cycle may not optimize fuel economy in real-world driving. Similarly, the CAFE system distinguishes between domestic and foreign automobile fleets, which can lead to distortions in the locations at which vehicles or parts are produced in ways that are not justified by market conditions.[10]

- **Mix Shifting.** In order to ensure compliance with an aggregate fleet-average standard, manufacturers may find it necessary to induce consumers to shift to smaller cars. As discussed in Chapter 3, such a trend may have safety consequences.

---

[8]As noted in Chapter 4, the production of a new vehicle involves the investment of energy resources that are equivalent to about two to three years of fuel use. Thus, the turnover of the automobile fleet has an energy cost, although much of the energy is not provided through petroleum. Evaluation of the impacts of fleet turnover should include consideration of the energy and environmental costs of production, as well as the energy and environmental savings that a more modern fleet would provide.

[9]This is not to suggest that actions by the manufacturers to use the regulatory system to best advantage are improper. Indeed, such behavior is consistent with the law and should be expected. In fact, it is the committee's view that the manufacturers are not alone in their gaming of the system. For example, it appears that at least part of the public's growing interest in light trucks and vans is a result of the more lenient CAFE and other requirements that apply to such vehicles. In effect, the public has exploited the system so as to avoid the effects of CAFE regulations.

[10]An automobile is deemed to be a foreign automobile unless 75 percent of its value added is of domestic origin (15 U.S.C. § 2003(b)(2)(E) (1988)). By adjusting the production location or the source of parts, a manufacturer has the capacity to allocate an automobile to either its domestic or its foreign fleet. Although the manufacturer must meet the CAFE requirements for each fleet, the distinction between foreign and domestic fleets does not limit or reduce the aggregate fuel economy of the manufacturer's total new car fleet. But, the distinction does have an impact on the location of production activities. For example, some manufacturers may choose to produce large cars abroad or to purchase parts for such cars abroad so as to "use" the CAFE opportunities for large-car sales that their foreign small-car production provides. Ford has turned to foreign sources of parts for its Crown Victoria for exactly this reason.

- **Costs and Benefits.** Fuel economy standards should be set at the point at which the marginal costs of the requirements balance the (overall) marginal benefits. Some evaluations have asserted that the costs of the CAFE standards far exceed their benefits (see, e.g., Charles River Associates, 1991; Leone and Parkinson, 1990; Shin, 1990). Others have found that the CAFE system is cost-effective (see Greene and Duleep, 1991). It seems, however, the CAFE approach, which involves the establishment of a fleet target by a political process many years in advance, will achieve a balance of costs and benefits only with difficulty.

Wholly apart from the cost-benefit balance, the costs of the system are largely concealed. The costs of the technologies to improve fuel economy are embedded in the prices of new cars. In addition, the system affects the pricing structure for cars: Full-line manufacturers may discount small cars and exact a premium for large cars so as to ensure a fleet mix that achieves compliance with the CAFE standard. The true costs are not easily tallied--a deficiency in any regulatory system.

Although the undesirable attributes of the CAFE system are significant, the approach has the following positive features:

- **Technology Forcing.** By imposing a mandatory fuel economy requirement for the fleet, the CAFE system forces manufacturers to introduce fuel-conserving technology. Although compliance with CAFE could be achieved by shifting the fleet mix to smaller cars, any such efforts would confront consumer resistance (at least at current fuel prices). Hence, there are strong incentives for the manufacturers to provide both fuel economy and vehicle characteristics that consumers favor. This direct technology-forcing aspect of CAFE is difficult to obtain through alternative approaches, particularly since the rational consumer may be relatively indifferent to significant variations in vehicle fuel economy (see Chapters 6 and 8).

- **Reduced Vulnerability to Increased Oil Prices.** The CAFE system does serve, in time, to bring about a more fuel-efficient fleet, an important and desirable goal. In effect, while oil is plentiful, it allows--perhaps, through lower per-mile operating costs, even encourages--the use of petroleum. A more efficient fleet also provides greater mobility when oil becomes scarce. Because it is impossible to turn over the fleet quickly, the establishment of a fuel-efficient fleet reduces (but does not eliminate) the vulnerability of the United States to the effects of an interruption in oil supplies.

- **Market Pressures.** The CAFE system reduces the capacity of foreign oil suppliers to raise prices. It also reduces current and potential demand for petroleum without decreasing the use of automobiles and light trucks. In effect, it diminishes the leverage that foreign petroleum cartels might otherwise be able to exploit.

Although the CAFE system has some favorable dimensions, the CAFE approach to achieving automotive fuel economy has defects that are sufficiently grievous to warrant careful reconsideration of the approach. Policymakers should explore other options or, if the continuation of the basic approach is deemed necessary, consider modifying the system to diminish its undesirable effects. The remainder of this chapter explores some of the alternatives that the committee believes warrant further scrutiny.

MARKET APPROACHES TO REDUCED FUEL CONSUMPTION

Various market incentives could be introduced to encourage consumer behavior that reduces fuel consumption. Two such approaches are discussed below: a policy of increasing the price of automotive fuel, and a system of fees and rebates to encourage the acquisition of fuel-efficient vehicles.[11] Both approaches have the advantage of using the market to achieve fuel economy by aligning consumer behavior with societal goals.

Increases in Fuel Price

One alternative--or supplement--to the existing CAFE system is a federal policy to increase the price of automotive fuel.[12] Federal efforts to enhance the fuel economy of vehicles reflect the conclusion that market forces are inadequate to ensure a usage of fuel that is consistent with societal objectives. Efforts to bring the price of fuel more in line with its "true" societal costs would allow market pressures to provide appropriate incentives to consumers.

Any strategy for increasing automotive fuel prices raises important issues of public and political acceptability. If an increase in fuel price were viewed as a tax, significant political costs would likely attend its enactment. Indeed, a gasoline tax can be regressive in that it has a differentially greater impact on people with lower income.

It is the committee's view that, properly considered, efforts to increase fuel prices should not be characterized as a tax. A gasoline price increase should be set so as to reflect the true cost of gasoline usage. Thus, a price increase is a price correction, not a tax purely for the purposes of raising revenue. Moreover, there are opportunities to collect fees at the gas pump relating to automotive usage that would otherwise be collected through other means. Such an approach would mean there could be no net

---

[11]Numerous other strategies could be considered. For example, incentives might be provided to encourage the drivers of older cars to purchase newer vehicles. Because older cars have poorer fuel economy than new cars, such a strategy raises the fuel economy of the fleet. Moreover, because older cars in general pollute the environment more than new cars, such a system would also have environmental benefits. However, despite recycling and especially extensive use of ferrous scrap, there are energy and environmental consequences from early retirement of vehicles, including more waste materials requiring proper disposal.

[12]The increase in price could be aimed at gasoline, at petroleum-based automotive fuels (gasoline and diesel), at petroleum, or even at carbon usage. A focus on automotive fuel would help reduce fuel consumption in the automotive sector, but it would not directly affect other parts of the economy. In contrast, a petroleum tax or a carbon tax would have widespread direct impacts throughout the economy and would achieve somewhat different objectives. A carbon tax, for example, presumably would reflect emphasis on responding to concerns about global warming. Given the scope of this report, the committee has focused on the implications of a strategy of increasing the price of petroleum-based automotive fuels. This does not imply that the committee rejects a more comprehensive strategy.

increase in payments by the average driver, but collection of such funds at the gas pump would provide incentives for improved fuel consumption. For example, there is widespread discussion of the nation's decaying transportation infrastructure--roads, bridges, and the like--and of the need for substantial investment in its restoration and improvement (National Research Council, 1987; U.S. Department of Transportation, 1991b). Funds collected at the gas pump for this purpose would pass the costs of this investment on to those who directly benefit from it.[13]

Indeed, there may be opportunities for investments of this kind that can reap improvements in fuel consumption and in safety that may exceed those attainable at equivalent cost through improved automotive technology. Such opportunities might include improved traffic management--improved road design, synchronized traffic lights, alternative-routing systems, and the like--and, in some circumstances, mass transportation systems. Such investments are important because a car in a gridlock is wasting fuel, no matter how impressive its CAFE rating. The trend toward increased numbers of vehicles on a road system that is not designed to accommodate them obviously compromises the achievement of efficient gasoline usage.

There are also possible opportunities to collect other types of fees at the fuel pump in a way that would not impose any net cost on the average consumer. For example, at least some portion of automotive insurance costs is properly related to the number of miles traveled and the location of that driving.[14] The recovery of a portion of the costs of insurance at the gasoline pump might allow an equitable recovery of insurance costs. Because insurance costs are a significant fraction of the operating costs of a vehicle, collecting a portion of them at the gas pump might also provide an important incentive for more efficient use of fuel with no net out-of-pocket impact on the average consumer.

In sum, the committee does not believe that political concerns about increased taxes should foreclose consideration of a strategy of increasing fuel prices. Indeed, efforts to increase the price of fuel would help to achieve the following important objectives:

- **Encouragement of Fuel Conservation**. Even if the CAFE system was not eliminated, an increase in fuel price would encourage conservation. The consumer would have incentives to acquire a fuel-efficient vehicle that is consistent with his or her needs *and* to use that vehicle in a way that conserves fuel. The existing CAFE system, in contrast, provides incentives to use improved automotive fuel efficiency at least in part in increased vehicle usage.

- **Impact on Entire Fleet**. An increase in fuel price would also provide incentives to improve the fuel economy of the entire vehicle fleet. Unlike the existing

---

[13]To the extent that any revenues collected at the pump are allocated to the improvement of the transportation system, the force of the claim that such increases in price have an undesirable regressive impact is reduced.

[14]Other types of fees--such as registration fees--might also be collected at the gas pump.

CAFE system, an increase in fuel price would affect all drivers (and the miles they choose to drive), not just the purchasers of new vehicles. And, to the extent that existing automobiles are not fuel efficient, an increase in fuel price would encourage turnover in the fleet.

- **Market Reinforcement of Societal Goals.** An increase in fuel price would encourage the consumer, within certain limits, to trade off vehicle attributes in a way that is consistent with societal objectives.[15] Unlike the existing CAFE system, the market would cause the consumer to weigh fuel economy more heavily among the attributes of the car (e.g., performance, size, weight) that are to be considered. Also, unlike the existing CAFE system, the market would provide incentives that reinforce societally desirable consumer decision making. The dissonance between the regulatory system and the market would be reduced.

- **Competitive Effects.** A policy of improving fuel economy by increasing fuel price would constrain full-line manufacturers (the domestic producers) in their ability to compete in various size classes only to the extent that they were unable to produce vehicles with the fuel economy of foreign manufacturers. (Increasing fuel prices would also cause mix shifts to small cars, thereby giving advantages to the strong producers in those segments.) There is no indication that the domestic manufacturers are unable to compete in this area.[16]

- **Market Efficiency.** Reliance on the price of fuel to encourage fuel economy in lieu of CAFE requirements would not affect the location of production operations. And, unlike the existing CAFE system, there would be less incentive to seek to "game" the regulatory system to exploit its defects. Perhaps even more important, if fuel prices were set at levels that fully reflected the externalities, the market in theory would serve to balance costs and benefits so as to ensure a cost-effective policy toward fuel efficiency.

Although reliance on increased fuel price in lieu of a CAFE system offers many advantages, the following drawbacks to the approach must also be considered:

- **Economic Disruption.** The size of the fuel price increase that would be necessary to achieve fuel savings equivalent to those demanded by aggressive CAFE standards might be quite substantial. The committee has had neither the time nor the capacity to evaluate fully the price increase for fuel that would induce reduced fuel consumption that is equivalent to that associated with aggressive CAFE standards. It may well be the case that the necessary increases in fuel price are so great as to make

---

[15]The environmental and safety attributes of automobiles are established as rigid requirements that all vehicles must satisfy. There would remain an important governmental role to ensure that these mandatory requirements reflect an appropriate balancing of societal objectives.

[16]In discussions with the committee, certain of the domestic manufacturers asserted that, when the size class of their vehicles is considered, they achieve a fuel economy that meets or exceeds that of their foreign competitors (see Chapter 5). Indeed, heightened consumer sensitivity to fuel economy might increase interest in 4-cylinder engines, a type of engine for which the domestic manufacturers currently have excess capacity.

sole reliance on a pricing strategy politically unacceptable because of the ripple effects on the economy of the adjustment to higher prices.[17] Moreover, such an increase in prices could upset reliance on low fuel costs, such as by those who live distant from their workplace.

- **Societal Disruption.** As discussed above, a reduction in vehicle miles traveled--one of the effects of fuel price increases--would reduce fuel consumption and diminish the environmental threat presented by automobiles. The achievement of those benefits would have its associated costs, however. American society is based on the mobility that low-cost fuel allows. If increasing fuel prices constrained travel significantly, there would be disruptive effects. Such a modification of American ways may be overdue, but one should not underestimate the implications to society of any reduction in mobility that might attend substantially increased fuel prices.[18]

- **Diminished Pressure for Technological Change.** Sole reliance on increased fuel prices to reduce fuel consumption would diminish the technology-forcing element of the existing CAFE system. Although the increased price of gasoline would create market pressures for the development of technology to improve fuel economy, the rational consumer would demand only those technologies that would pay for themselves through fuel savings. Moreover, only a few of the available technologies that might be demanded by an aggressive CAFE standard may be cost-effective unless very substantial increases in fuel price occurred.[19] The CAFE system, in contrast, encourages technological advances that do not pay for themselves in terms of fuel savings, even at elevated fuel prices.[20] Although a rational consumer might not choose

---

[17]A report prepared for the Motor Vehicle Manufacturers Association by Charles River Associates (1991) asserts that a gasoline tax of 27.5 cents per gallon that is phased in at an annual rate of 5.5 cents per gallon from 1992 to 1996 and held constant thereafter would yield, at less expense, cumulative gasoline savings until 2002 exceeding those from a CAFE standard that dictates 40-mpg fleet averages by the year 2001. These excess savings accrue over the short term because the gasoline tax immediately affects all vehicles, including the new car fleet. However, because the impacts of the CAFE system are delayed by reason of the diluted impacts of new car fuel efficiency on fleet-wide fuel efficiency, the long-term effects of a 40-mpg CAFE standard would be much more substantial than the 27.5 cents per gallon fuel tax. Other analyses suggest that much greater price increases (perhaps in excess of $1 per gallon) would be required to achieve the impact of such CAFE requirements (U.S. Department of Transportation, 1991a).

[18]Increased fuel prices might encourage drivers to reduce miles traveled in ways that do not require a significant modification of life-style. For example, shoppers might plan their trips more carefully and workers might join carpools or use public transportation. The unavoidable impacts of increased fuel costs might fall most heavily on the rural poor.

[19]Although other countries have fuel prices substantially higher than of those in the United States, this fact has not induced the introduction of new technologies offering radically enhanced fuel economy. Fuel consumption in those countries is lower on a per-vehicle basis because of the types of automobiles that are purchased--typically smaller cars with manual transmissions and without air conditioning--and because people drive less.

[20]Moreover, a CAFE approach can spur the introduction of even cost-justified technology at a faster pace than the market. In order to ensure the reliability of new technology and to prevent the premature obsolescence of current production facilities, a manufacturer typically introduces a new technology in only a small portion of its fleet. The experience that is gained can reduce the risk associated with the widespread introduction of the

such a vehicle in the market,[21] the CAFE approach creates an inventory of fuel-efficient automobiles, which provides some protective capacity in the event of a petroleum interruption.

- **Mix Shifting.** Significantly increased fuel prices would induce a reduction in miles traveled and mix shifting by consumers to smaller cars. Thus, although the safety impacts of future downsizing may be uncertain, any safety effects brought about by aggressive CAFE requirements might also be associated with a strategy of sharply increased fuel prices. The safety impacts would be reduced, however, because the impacts of mix shifting would be counteracted to some extent by reduced miles traveled and by the trading-in of older, less safe cars for new cars with improved safety.

- **Impacts on Manufacturers.** An increase in the price of fuel, if not collected in some revenue-neutral way, would increase the cost of owning and operating a vehicle. If the price increase was significant, aggregate demand for new vehicles would be reduced, which would have adverse impacts on the auto industry and its employees. Moreover, as noted above, increased fuel prices could induce mix shifting by consumers, which would have a differential effect on the various manufacturers. Nonetheless, in their presentations to the committee, the automobile manufacturers have uniformly identified a strategy of increasing fuel prices as a desirable alternative to the CAFE system.

The committee recognizes that balancing the costs and benefits of a strategy of using fuel prices to improve fuel economy in lieu of CAFE-style regulations involves a significant measure of judgment on which reasonable people may well differ. The committee nonetheless believes that, at the least, a strategy of *complementing* fuel efficiency regulations with modest fuel price increases may be appropriate so as to minimize at least some of the disadvantages associated with CAFE standards. For example, even limited price increases for fuel would reduce the gulf between market signals and the regulatory requirements.

Although conscious of the delicate state of the economy, the committee believes that market circumstances warrant consideration of a policy of increasing fuel prices. As shown earlier (refer to Figure 1-4), the price of gasoline is now lower in real terms than at any time in the recent past. In such circumstances, an increase in the price of gasoline would, in part, *restore* price levels. Indeed, because of the low real price of gasoline, the CAFE system is increasingly inconsistent with market pressures, which exacerbates the various problems described above.

---

technology. An aggressive CAFE system could force the manufacturers to accept increased risk and cost in introducing technology.

[21]Indeed, as shown in Chapter 6, a rational consumer might be relatively indifferent to fuel economy ratings over a considerable range.

The United States stands alone among the industrial nations in its policy toward gasoline prices. As shown in Figure 9-1, the price of gasoline in other countries is significantly higher than in the United States. (The increments above U.S. prices are largely the result of taxes, not of higher costs of production.) Other nations have been able to absorb increased fuel prices without significant adverse impacts, and the committee believes that the United States might also accommodate an increase in price. Although the United States may be unique in the degree to which its citizenry relies on the automobile for personal transportation, this fact should not inhibit a pricing strategy for fuel that reflects a larger share of its true costs.[22]

On balance, the benefits of a policy of increasing fuel prices seem to outweigh the disadvantages. Whether sole reliance should be placed on increased fuel prices as a means to improve fuel economy presents issues that require further scrutiny. Serious consideration should be given to such matters.

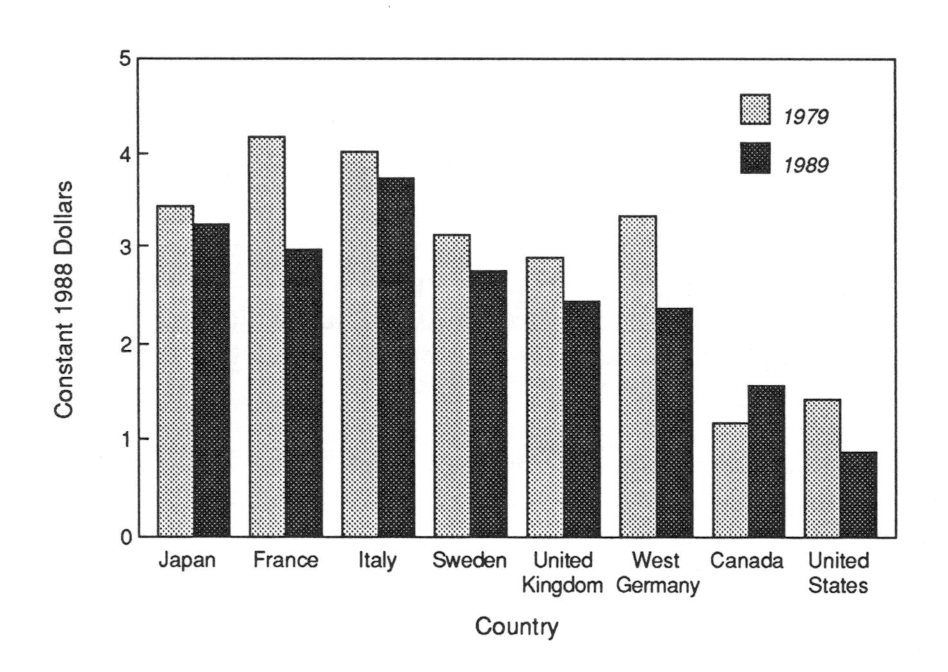

FIGURE 9-1 Gasoline prices for selected countries, 1979 and 1989.

SOURCE: Oak Ridge National Laboratory (1991).

[22]An increase in gasoline prices may also commend itself as a national policy for other reasons. The recent Persian Gulf war has reemphasized the vulnerability of the world's petroleum supply. An increase in fuel prices would not only reinforce desirable consumer behavior, but could supplement the nation's Strategic Petroleum Reserve in that it would provide a buffer to the shock to the economy that could arise from a severe supply disruption. A supply disruption could threaten a sharp run-up in gasoline prices and, in such circumstances, the government could reduce the surcharge on gasoline so as to reduce the resulting economic impacts.

Fees and Rebates for Fuel Economy

Other market-based strategies to encourage reduced fuel consumption are perhaps less contentious than increases in fuel prices. For example, to induce consumers to purchase vehicles with above-average fuel economy, a system of fees and rebates might be established. Consumers who purchase a vehicle with below-average fuel economy could be charged a fee and consumers who purchase a vehicle with above-average efficiency could be given a rebate.[23] Any such system could be revenue neutral; the structure of the fees could be designed to cover the rebates.

The "feebate" system could be applied in combination with or in place of CAFE-style regulation or increased fuel prices. If applied by itself, the feebate approach would eliminate some of the disadvantages of the CAFE approach. There would be pressure on the manufacturers to produce vehicles with higher fuel economy, but the pressure would not be at odds with the demands of consumers. In effect, the feebate system would allow manufacturers to produce whatever consumers choose to buy; the feebates would channel consumer decision-making in a societally desirable direction. Moreover, there would be no need to establish a fuel economy standard because the feebate would exert continuing pressure for increased fuel economy. Although full-line producers would be disadvantaged in any such system--their large-car offerings would become more expensive--the system would not provide the same incentives as the CAFE approach for small-car manufacturers to intrude into the large-car market. Thus, at least some of the perverse competitive impacts of the CAFE system would be muted. And, unlike the CAFE approach, the costs of the regulatory system would be largely transparent, which enhances sound societal policymaking.

The feebate approach also has its disadvantages. Like the CAFE system, feebates would affect only the new car fleet and, in fact, could discourage the retirement of fuel-inefficient and environmentally objectionable large cars. Moreover, like the CAFE system, it would not provide any incentives to reduce the number of miles traveled; indeed, some of the gains in fuel economy would be offset by increased travel. And, if the feebates were substantial, they could induce mix shifting by consumers, which could have safety consequences.

A feebate system is unlikely to have the technology-forcing effect of a CAFE system unless the size of the fees and rebates became very substantial. The system would induce manufacturers and consumers to compare the costs of the technology to achieve a given improvement in fuel economy with the benefits--the resulting change in the fee or rebate and the present value of the reduced operating costs for fuel. Rational manufacturers would incorporate those technologies that "pay for themselves" in this way. The system might induce less extensive introduction of fuel economy technology than an aggressive CAFE requirement, but, on the other hand, the cost

---

[23]There are a variety of ways to structure a feebate system. For example, the fee might be established as $10 for every gallon of gasoline used over a typical year in excess of what would be used by an average car, and the rebate could be calculated in terms of the gasoline use avoided. If the average car got 30 mpg, the owner of a car that achieves 20 mpg would confront a fee of approximately $1,700, whereas the owner of a 40-mpg car would receive a subsidy of about $800, based on 10,000 VMT.

ceiling established by feebates would be a reasonable and practical constraint on investments in technologies that are not economically justified.

IMPROVING THE CAFE SYSTEM

Even if it is concluded that market-based alternatives to CAFE regulations are infeasible, consideration should be given to modifying of various aspects of the CAFE system.

Timing

Because of the nature of the automotive development process, extensive lead time is required to prepare for the introduction of a new automobile. At the time this report was written, the engineering design for cars for MY 1996 was largely fixed. There is no reasonable way to impose significantly enhanced fuel economy requirements on vehicles that are three to four years away from sale without imposing very substantial costs--either very expensive retrofits or price increases to force a mix change. In light of this fact, there is little opportunity by MY 1996 to achieve fuel economy gains that are substantially in excess of those that are already in the product plans of the manufacturers. Any modifications of the CAFE system must recognize this fact.

As the time horizon moves to the more distant future, there is an opportunity to establish more significant fuel economy requirements. By establishing firm requirements with an adequate lead time, the regulatory system provides the manufacturers with the opportunity to plan for change in an economically efficient way. For example, the need to achieve certain levels of fuel economy could be accommodated by introducing a new engine that incorporates certain technology at the time that an older engine would have been replaced in any event. The difficulty with establishing requirements in the very distant future, however, is uncertainty: It is difficult to establish standards that will challenge the manufacturers, but that are also practically achievable. Establishing such standards requires significant elements of judgment, and the stakes are very substantial if the judgment is erroneous.

Any modification of the CAFE standards should be guided by these realities. Limits for vehicles to be produced in five years (1996) should be substantially similar to the fuel economy already planned by the manufacturers. Standards for vehicles to be produced in 10 years (2001) might be fixed so as to establish a firm planning target for the manufacturers. Standards for cars to be produced in 15 years (2006) might also be established so as to guide advanced product planning and cause the manufacturers to conduct the necessary advanced engineering and research. But, there must be some measure of flexibility in such a distant target to accommodate uncertainty. The 15-year target could then be reconsidered after 5 years and a firm requirement established for vehicles that are then to be produced in 10 years. Such a rolling system of standards satisfies the need for certainty, but also recognizes the need for flexibility to accommodate the uncertainty in distant estimates.

Alternatives to Uniform Fleet Targets

Under the existing CAFE system each manufacturer is required to achieve a specified fleet fuel economy. As discussed above, the approach imposes greater burdens on manufacturers that produce predominantly large cars than on those that produce small cars. This aspect of the CAFE system has disadvantaged U.S. and some European manufacturers in comparison with Japanese manufacturers.

Legislation under consideration in Congress (S. 279) would address this problem by requiring certain mandatory percentage gains in CAFE performance ratings by each manufacturer over the level that was attained in a specified base year. This approach would establish different requirements for different manufacturers and would have the perverse effect of requiring those manufacturers with the best fleet fuel economy in the base year to comply with CAFE requirements in the outlying years that are more stringent than those for manufacturers who had not achieved similar accomplishments. The regulatory system would thus penalize those manufacturers who have exceeded the minimal requirements and thereby discourage any fuel economy accomplishments above the baseline in the future. Moreover, the approach is unfair because the currently available technology for improving fuel economy might already have been incorporated in the base year by the manufacturer who is confronted with the largest future-year fuel economy requirements. In addition, the selection of the base year could create arbitrary advantages or disadvantages for the manufacturers based on the happenstance of the product mix or technology that was applied by the manufacturers in that year.

The percentage approach would also have important differential effects among the manufacturers. As discussed above, the existing CAFE system has operated to the disadvantage of the full-line manufacturers (the domestic industry) and to the advantage of the producers of small cars (particularly the Japanese). The percentage approach would reverse this effect. Because the small-car manufacturers--who would have high base-year CAFE accomplishments--would be required to achieve higher absolute CAFE requirements under the percentage approach than full-line manufacturers, the small-car manufacturers would have a particular challenge in changing their product mix to include a higher percentage of large cars. As discussed in Chapter 5, the Japanese manufacturers in recent years have sought an increasing share of the large- and luxury-car market. The percentage approach would protect domestic manufacturers from this competition.[24]

A variety of other alternatives to the current CAFE scheme are worthy of consideration. A revised system might establish fuel economy requirements that are tailored to an attribute of the car, such as usable passenger volume (or passenger-carrying capability) or perhaps size class. For example, cars with larger volume might be required to meet lower fuel economy standards than smaller cars, in recognition that

---

[24]Some Japanese auto manufacturers have claimed that the percentage approach is protectionist and constitutes an illegal trade barrier (comments of American Suzuki Motor Corporation on S.1224, U.S. Congress, 1989).

the achievement of equivalent fuel economy is more difficult in the larger car. The standards would have to be set, however, so as to challenge the manufacturers to improve the fuel economy of all their vehicles.[25] This approach has an advantage over the existing system in that it could be applied so as to ensure that each manufacturer is confronted with a roughly equal challenge regardless of the differences in the sizes of cars that the manufacturer produces. If an attribute or class standard is not imposed on cars and trucks, then a relatively lower CAFE standard may have to be formulated in order to minimize the adverse consequences for full-line domestic manufacturers.

Unlike the current CAFE system or the percentage approach, a revised system based on a vehicle attribute does not ensure that a given overall fleet fuel efficiency will in fact be achieved. Manufacturers would be under less pressure to adopt a pricing structure that ensures the sale of sufficient small cars so as to achieve compliance with the fleet standard.[26] The attribute-based system thus would not provide any incentive for the manufacturers to encourage mix shifting to small, fuel efficient vehicles. Such a change may be a necessary cost, however, of a system that does not disadvantage the full-line manufacturers and that creates challenges for the enhancement of fuel economy by all manufacturers, regardless of the type of cars they produce.

The selection of the attribute (or attributes) that would provide the basis for the revised system admittedly has an element of arbitrariness. For passenger cars, interior volume (or passenger-carrying capacity) appears to be a sensible criterion because it is obviously related to consumer vehicle use. Some people, however, may buy large cars despite an absence of real need for their carrying capacity, and the system provides no incentives for such a consumer to make another choice. Moreover, attributes other than volume could be suggested. For example, performance is an attribute demanded by consumers and its satisfaction, like an increase in volume, tends to undermine fuel economy. Some might suggest that the fuel economy standard could be adjusted to allow higher performing cars to benefit from more lenient standards than lower performing cars. Performance has a very different character than interior volume, however, because its social utility, once certain levels have been achieved, is more debatable. Nonetheless, the selection of the attribute (or attributes) on which to base the system would raise questions of judgment.

Another difficulty with an attribute-based approach is that it might be susceptible to gaming. For example, a volume-based standard might provide incentives for a manufacturer to increase the volume of vehicles regardless of whether the increased volume enhances utility for drivers.[27] Similarly, a standard based on size class would

---

[25] This approach might be combined with a corporate-averaging scheme so as not to require the wholesale modification of every model in a given year (Office of Technology Assessment, 1991).

[26] Under the existing system, small cars may be more attractively priced than larger cars so as to ensure that the fleet CAFE standard is achieved. This may have desirable social consequences in that it makes vehicles more available for less wealthy purchasers than would otherwise be the case.

[27] Chrysler Corporation presentation to the committee's Subgroup on Standards and Regulations, November 4, 1991, Washington, D.C.

provide an incentive for the manufacturer of a vehicle near the boundary of a class to add additional volume to gain the benefit of the lower fuel economy requirement of the larger size class. The opportunities for gaming and the means, if any, for reducing undesirable side-effects warrant careful scrutiny. The committee has not had an opportunity to pursue such issues in depth, but it believes they require careful examination.

Light Trucks

As discussed above, light trucks have grown to nearly 30 percent of the total fleet. Currently, these vehicles must comply with significantly more lenient CAFE requirements than automobiles,[28] with the consequence that the aggregate fuel economy of the vehicle fleet is compromised by the growth in the light-truck segment. Although trucks and vans used for load-carrying or towing functions should have lower fuel economy than cars due to engineering constraints, the evidence suggests that most consumers of certain popular categories of light trucks use these capabilities only occasionally.[29] Moreover, the manufacturers have not been required to achieve fuel economy improvements for light trucks that are commensurate with those required of automobiles. The percentage improvement in the fuel economy of light trucks since 1973 has been about three-fifths of that attained by automobiles (see Chapter 1). The regulatory system should better serve to integrate and conform the requirements for light trucks with those for automobiles.

The precise means for achieving such conformance requires further examination. One approach is to include certain categories of "trucks"--small vans, light pickups, and sport utilities--in the automobile CAFE category in recognition of the fact that these vehicles are largely used in the same way as automobiles.[30] (Some corresponding adjustment of the aggregate automobile CAFE standard would have to be made to reflect the legitimate constraints on the revised fleet.) But, such an approach would have an important differential effect among the various manufacturers because of their different shares of the light-truck market. Chrysler, for example, is a significant producer of vans and could be seriously affected by such a change.

Domestic Content

The existing CAFE system requires that each manufacturer separately satisfy the CAFE requirements for its domestic and its foreign fleet. The distinction was introduced so as to ensure that manufacturers currently producing large cars in the United States would have an incentive to produce small cars here as well. Although

---

[28]Whereas automobiles must achieve a fuel economy rating of 27.5 mpg in 1991, light trucks need only achieve 20.2 mpg. 49 C.F.R. §§ 531.5, 533.3 (1990).

[29]Presentation to the committee at the July 8-12, 1991, workshop in Irvine, Calif., by the Bureau of the Census entitled "Census of Transportation: The Truck Inventory and Use Survey."

[30]Similar approaches could be designed for various alternative CAFE systems. See, e.g., Office of Technology Assessment (1991:104).

the approach may serve that end, it also provides incentives to produce large cars abroad (or to keep domestic content below the 75 percent limit) so as to avoid wasting any CAFE benefits that are gained from the production of small cars abroad. Indeed, it appears that there is a substantial measure of arbitrariness in the way the system is applied because of the complications associated with evaluating foreign and domestic content. (Business Week, 1991; Fortune, 1991; Wall Street Journal, 1991).

It is the committee's view that the domestic-content provision has no obvious or necessary connection to the achievement of fuel economy. Moreover, it is not clear that the restriction achieves any net gain in the domestic manufacture of cars or the preservation of American jobs. Although it may encourage the production of small cars in the United States, it also encourages the production of large cars abroad. The limitation should be evaluated to assess whether the restraint it places on the market achieves any net benefits. Consideration should be given to the elimination of this provision.

CAFE Credits

The fleet fuel economy achieved by a manufacturer is not completely within the manufacturer's control: It is governed by the mix of vehicles that consumers decide to purchase.[31] In light of this fact, it is important that the system provide sufficient flexibility so as to allow the manufacturers to avoid a penalty for shortfalls that are not persistent from year to year. The existing CAFE system provides a measure of flexibility by allowing the carry-forward and carry-back of "credits" for fleet fuel economy that exceeds the CAFE requirements. In this way, manufacturers can avoid the penalty that would otherwise accrue for a temporary shortfall in any given year, at least to the extent that it is "covered" by credits in other nearby years. But, perhaps greater flexibility should be allowed.

The committee believes that there may be some merit in allowing the manufacturers to trade CAFE credits. That is, a manufacturer that exceeds CAFE requirements might be allowed to sell credits to manufacturers that fall below the standards. In effect, such trading is currently going on through the sale by some manufacturers under their own marque of vehicles produced by other manufacturers. (This practice is termed *rebadging*.) Although direct trading of credits may have differential impacts on foreign and domestic manufacturers, it should be considered because it enhances economic efficiency.

Unlawful Conduct

The existing law provides a penalty of $5 per car for each 0.1 mpg by which a manufacturer falls short of its fleet CAFE requirements (15 U.S.C. § 2008(b)(1) (1988)). Because a shortfall on CAFE compliance constitutes unlawful conduct (15 U.S.C. § 2007 (1988)), however, the violation of the CAFE limit is not a socially

---

[31] Within certain limits the manufacturers can use price as a means to channel consumer decisions. Nonetheless, the mix of cars that are sold is not a variable that is entirely subject to a manufacturer's control.

or legally neutral act. As a result, the domestic manufacturers have stated that any plan to depart from the CAFE approach presents the threat of legal exposure in addition to the penalties. (The foreign manufacturers do not have the same inhibitions; some foreign manufacturers appear to view the payment of the penalty as a cost of doing business.) The system might be modified so as to remove the characterization of a CAFE violation as unlawful conduct, thereby allowing the manufacturers to trade-off the costs of compliance with the penalty.[32] Such a modification would put an effective cap on the cost of compliance with the CAFE requirements--thereby enabling some balancing of costs and benefits, a real advantage. Such a modification of the system, although subtle, would effectively introduce some of the elements of the feebate system within the existing regulatory structure.

OTHER POLICIES FOR REDUCING FUEL CONSUMPTION

As noted above, the chief purpose of the legislation establishing fuel economy standards was to reduce the consumption of petroleum by cars and light trucks. If this objective is to be served, it is important to recognize that a variety of approaches in addition to those discussed above can have an important impact on petroleum use. An appropriate strategy for reducing petroleum consumption is to consider a variety of tools to address the problem and to select those that can achieve the objective at the lowest cost. An appropriate strategy need not rely on one approach--for example, reliance on CAFE standards--to achieve the policy objective.

Although the committee did not examine other approaches in detail, a number are mentioned below because they touch in important ways on fuel economy:

- **Infrastructure Improvements.** Real-world fuel economy is directly governed by the capacity of the road system to enable drivers to travel quickly, efficiently, and safely. As noted above, the nation suffers from a decaying transportation infrastructure. Improvement of roads, bridges, highways, and the like would not only improve safety, but also offer fuel economy benefits by improving traffic flow and reducing delays (National Research Council, 1987).

- **Intelligent Vehicle-Highway Systems.** Information and computer technology can be harnessed to improve traffic flow. There are numerous opportunities to do so through providing alternate route and road information, synchronized traffic lights, and the like (National Research Council, 1991).

- **Public Transportation Systems.** Public transportation systems can provide the means for the movement of large numbers of people at very low consumption of fuel per passenger mile. Indeed, electrically powered public transit may not be dependent on petroleum at all, although it still may contribute indirectly to greenhouse gas emissions because of the use of fossil fuels in the generation of electricity. In

---

[32]The modification might be accompanied by an adjustment of the fee. The fee should be set at a level that approximates the social costs of noncompliance.

appropriate circumstances, improved public transit can be a means for reducing demand for petroleum and for limiting the traffic load on congested streets.

- **Highway Speeds**. As discussed above, the aerodynamic drag on a vehicle is proportional to the square of the velocity of the vehicle. As a result, a vehicle traveling at 55 mph can operate significantly more efficiently than one traveling at 65 mph (or higher). The reduction of speed limits, although politically unpopular, is a rational strategy for reducing petroleum consumption and enhancing safety (National Research Council, 1984).

- **Car Pooling**. A significant portion of fuel consumption is related to commuting to and from work. Tax or other policies could be adjusted so as to encourage car pooling and thereby reduce the number of vehicles involved in commuting.

Although this study has focused on the means for enhancing the fuel economy of cars and light trucks, the objective of reducing the consumption of petroleum can be achieved by a wider set of policy tools than has been examined. All such policies should be weighed and applied as appropriate.

SUMMARY

The means by which improved fuel economy is attained have important implications for consumers, manufacturers, and automotive workers. This review leads to the following observations:

- The exiting CAFE system has defects that warrant careful examination. Chief among the defects is the fact that the CAFE system, over recent years, has been increasingly at odds with market signals, which mutes and diminishes the system's effectiveness and increases its costs. Moreover, the CAFE system has operated to disadvantage the domestic manufacturers in competition with certain foreign manufacturers (principally Japanese).

- A policy of increasing fuel prices warrants consideration either as an alternative to fuel economy regulation or as a complement to it. Increasing fuel prices would internalize the costs associated with fuel consumption and would thereby provide appropriate market signals to channel consumer behavior in a direction consistent with societal objectives.

- A system of fees and rebates that are related to the fuel economy of vehicles might also be considered. Such "feebates" would encourage the acquisition of fuel-efficient vehicles.

- In determining whether the CAFE system should be retained, Congress should consider several modifications, including the following:

-- Establish a rolling set of requirements that reflect the need to provide the manufacturers with adequate lead time to comply with CAFE requirements, but that allow for modification of requirements that are distant in time.

-- Alter the system so as to present equivalent technical challenges to manufacturers, regardless of the type of vehicles they produce. Standards based on a relevant vehicle attribute, such as interior volume, should be examined.

-- Bring the fuel economy standards for light trucks into conformance with those for automobiles.

-- Study the elimination of the distinction between domestic and foreign fleets for CAFE purposes.

-- Enhance the flexibility of the CAFE system by allowing the trading of CAFE credits.

-- Modify the law so as to remove the characterization of noncompliance with a CAFE limit as unlawful conduct and increase the penalty to a level that approximates the social cost of noncompliance. This change would increase the flexibility to respond to the law in an economically efficient way.

- The percentage-improvement approach to CAFE regulation--an approach proposed in legislation pending in Congress--has the perverse effect of requiring those manufacturers with the best fleet fuel economy in a base year to comply with CAFE requirements in the outlying years that are more stringent than those applied to manufacturers with lesser accomplishments. The selection of the base year would create arbitrary advantages and disadvantages, depending on the happenstance of the product mix or technology that was applied by the manufacturers in the base year. The system is unfair because the manufacturers confronting the most stringent requirements may already have incorporated in the base year many of the available technologies for improving fuel economy. Moreover, the system would limit efforts by the Japanese manufacturers to increase their market share of larger cars and thereby reduce competition in that segment.

- The objective of reducing petroleum consumption can be achieved by a variety of means. In addition to the CAFE system (and its variants) and increased fuel prices, the available policy instruments include improvements in the transportation infrastructure, intelligent vehicle-highway systems, improved public transit, reduced speed limits, and incentives for car pooling. All such policy instruments should be considered in developing an appropriate federal strategy for reducing petroleum consumption.

REFERENCES

Broadman, H. J. 1986. The social cost of imported oil. *Energy Policy* June:242-252.

Bureau of the Census. 1989. *Projections of the Population of the United States, by Age, Sex, and Race: 1988 to 2080.* Current Population Reports, Series P-25, No. 1014. Washington, D.C.: U.S. Government Printing Office.

Business Week. 1991. Honda: Is it an American car? November 18:105.

Consumer Attitude Research (C.A.R.). 1991a. *C.A.R. Report*. 2nd qtr. Birmingham, Mich.

Consumer Attitude Research (C.A.R.). 1991b. *Reasons for Buying*. Brimingham, Mich.

Charles River Associates. 1991. *Policy Alternatives for Reducing Petroleum Use and Greenhouse Gas Emissions*. Boston, Mass.

Dahl, C. 1986. Gasoline demand survey. *Energy Journal* 7:67-82.

DeCicco, J.M. 1991. Cost-effectiveness of fuel economy improvements. Paper presented at the workshop of the Committee on Fuel Economy of Automobiles and Light Trucks, Irvine, Calif., July 8-12. American Council for an Energy-Efficient Economy, Washington, D.C.

Difiglio, C., K.G. Duleep, and D. Greene. 1990. Cost effectiveness of future fuel economy improvements. *Energy Journal* 11(1):65-86.

Fortune. 1991. Do you know where your car was made? June 17:52.

Greene, D.L. In press. Vehicle use and fuel economy: How big is the rebound effect? *Energy Journal*, forthcoming.

Greene, D.L., and K.G. Duleep. 1991. *Costs and Benefits of Automotive Fuel Economy Improvement: A Partial Analysis*. Tenn.: Oak Ridge National Laboratory.

Greene. D., and J.N. Liu. 1988. Automotive fuel economy improvements and consumers' surplus. *Transportation Research* 22:203-218.

Kahn, J.A. 1986. Gasoline prices and the used automobile market: A rational expectations asset price approach. *Quarterly Journal of Economics* May:323-339.

Leone, R.A., and T.W. Parkinson. 1990. *Conserving Energy: Is There a Better Way? A Study of Corporate Average Fuel Economy Regulation*. Washington, D.C.: Association of International Automobile Manufacturers.

Motor Vehicle Manufacturers Association. 1991. *Facts & Figures '91.* Washington, D.C.

National Research Council (NRC). 1984. *55: A Decade of Experience.* Transportation Research Board Special Report 204. Washington, D.C.: National Academy Press.

National Research Council (NRC). 1987. *Infrastructure for the 21st Century--Framework for a Research Agenda.* Committee on Infrastructure Innovation. Washington, D.C.: National Academy Press.

National Research Council (NRC). 1991. *Advanced Vehicle and Highway Technologies.* Committee to Assess Advanced Vehicle and Highway Technologies. Washington, D.C.: National Academy Press.

Oak Ridge National Laboratory. 1991. *Transportation Energy Data Book.* 11th ed. ORNL-6649. Tenn.

Office of Technology Assessment, U.S.Congress. 1991. *Improving Automobile Fuel Economy: New Standards, New Approaches.* Washington, D.C.: U.S. Government Printing Office.

Shin, D. 1990. *The Costs and Benefits of Federally Mandated Policies to Promote Energy Conservation: The Case of the Automobile Efficiency Standard.* Research Study No. 050. Washington, D.C.: American Petroleum Institute.

U.S. Department of Transportation. 1991a. *Auto Industry Comments on: Briefing Book on the United States Motor Vehicle Industry and Market, Version 1.* Cambridge, Mass.: John A. Volpe National Transportation Systems Center.

U.S. Department of Transportation. 1991b. *Briefing Book on the United States Motor Vehicle Industry and Market*, Version 1. Cambridge, Mass.: John A. Volpe National Transportation Systems Center.

U.S. Department of Transportation. 1991c. *The Status of the Nation's Highways and Bridges: Conditions and Performance.* Report of the Secretary of Transportation to the U.S. Congress. Washington, D.C.: U.S. Government Printing Office.

U.S. Congress, Senate. 1989. Hearings on S. 1224, The Motor Vehicle Fuel Efficiency Act of 1989. Consumer Subcommittee of the Commerce, Science and Transportation Committee. Sept. 7.

Wall Street Journal. 1991. Foreign or domestic car firms play games with the categories. November 11.

APPENDIX A

PRESS RELEASE ANNOUNCING FUEL ECONOMY STUDY

December 26, 1990
FOR GENERAL RELEASE

National Research Council
COMMISSION ON ENGINEERING AND TECHNICAL SYSTEMS
ENERGY ENGINEERING BOARD

AN EVALUATION OF THE POTENTIAL AND PROSPECTS
FOR IMPROVING THE FUEL ECONOMY OF NEW AUTOMOBILES
AND LIGHT TRUCKS IN THE UNITED STATES

INTRODUCTION

The purpose of this study is to estimate fuel economy levels that could practically be achieved in new automobiles and light trucks (up to 8500 lbs. gross vehicle weight rating) produced for the United States market in the next decade.

The study has been requested by the National Highway Traffic Safety Administration to ascertain the potential and prospects to improve the fuel economy of new vehicles, while meeting existing and pending environmental and safety standards for the vehicles.

The study will be conducted in two phases. The work under Phase 1 is to be completed by June 30, 1991 and that under Phase 2 by March 31, 1992.

OBJECTIVES

Phase 1 of the study is expected to provide, on a "best judgment" basis, estimates by size class of vehicles (e.g., full-sized, mid-sized, compact, and sub-compact passenger cars, and large and small light trucks) produced by automotive corporations with major assembly facilities in the United States and Canada of fuel economy practically achievable in the next decade, taking into consideration, as appropriate, provisions of the Clean Air Act Amendments of 1990, the state of the art in the applications of technologies relevant to achieving higher fuel economy and improving safety, and the viability of the domestic automotive industry in the U.S. market. Phase 1 work is also expected to result in the identification of principal barriers in the United States that appear to constrain the rates at which technologies enhancing fuel economy can be introduced and sustained in the marketplace.

Phase 2 of the study will analyze alternative measures to overcome the principal barriers to the technologies considered in Phase 1.

PROPOSED EFFORT

A committee will be appointed by the National Research Council to carry out this study. People with requisite qualifications will be sought for membership on the study committee with expertise in areas such as the following: internal combustion engines, fuels and lubricants, drive trains, automotive structures and materials, emission control systems, vehicle design, manufacturing of cars and light trucks, safety, financial practices and markets relevant to the automotive industry, federal and state regulations under which the automotive industry functions, consumer behavior, and automotive industry/U.S. economy interactions. A committee slate will be sought that is balanced with regard to the science and technology type of credentials and those from other disciplinary areas such as finance, economics, regulations, and behavioral sciences.

In Phase 1, the Committee will rely primarily on mechanisms such as the following to expeditiously obtain information pertinent to the study:

(a) The Committee will invite structured presentations, to be delivered at committee meetings and in a workshop forum, from domestic and foreign automobile manufacturers and their suppliers; from representatives of qualified organizations closely involved with but functioning outside of the automotive industry per se; from the National Highway Traffic Safety Administration and its contractors and subcontractors as appropriate; and from other relevant parties (individuals, firms and other entities in the private sector, and government agencies).

(b) The Committee will avail itself of the data and analytical resources of the National Highway Traffic Safety Administration that would be relevant to the study including, as appropriate, the resources of the National Highway Traffic Safety Administration's contractors and subcontractors who specialize in studies of the automotive industry and markets. The National Highway Traffic Safety Administration will facilitate the Committee's use of these resources.

(c) The Committee will commission expert written reviews of selected topics from the extant literature, for example, trade-offs in automotive design of weight versus safety; dynamics of automotive industry changes since the Arab Oil Embargo of 1973; myths and realities in consumers preferences for automobiles; and so forth.

On the strength of what the Committee ascertains from the foregoing processes, the following tasks will be addressed:

PHASE 1

Task 1

The Committee will evaluate technologies in conventionally powered cars and light trucks that could, in the time frame of the next decade, contribute to improved fuel economy of new vehicles. Examples of technologies that might be presented to the Committee include the following: front-wheel drive; reductions in aerodynamic drag; 4-and 5-speed automatic transmissions; torque converter lockup; electronic and computer controls; continuously variable transmissions; 6-speed manual transmissions; high efficiency accessories; electric power steering; engine improvements (e.g., from components design, controls, materials); 2-cycle engines; diesel engines; improved lubricants; energy storage; reductions in rolling resistance and other driveline losses; weight reductions; reductions in horsepower-to-weight ratios.

In its evaluation, the Committee will consider factors such as the following:

a. The magnitude of fuel economy improvements that can be expected from the technologies, singly or in combinations.

b. The time at which the technologies could be introduced and the rates at which they might penetrate the U.S. market, given existing industrial capabilities in the United States and limitations (e.g., technical, financial, regulatory, organizational, and marketing limitations) to deploying improved or new capabilities in the next decade.

c. Likely effects in the United States of the technologies on initial and life-cycle costs of vehicles and vehicle safety, taking account of the effects on fuel economy of the interaction between and among technologies.

For the purposes of evaluation, the Committee will consider defining a baseline with vehicle size, size mix, equipment and performance consistent with the 1990 model year new cars and light trucks sold in the United States. Measures of fuel economy will be based on the EPA Test Cycle, and assumptions regarding future automotive fuel prices may be based on projections made by the Department of Energy and other sources of such projections available in the public domain.

Task 2

The Committee will identify and describe the principal barriers to the introduction in the United States of the technologies underlying the improvements in the fuel economy of new vehicles.

In performing this task, the committee will use information presented by the automobile manufacturers, by the National Highway Traffic Safety Administration, and by others. Such information is expected to cover topics such as the following: manufacturers' capital and operating costs in aggregate; research and development plans and costs, technology development and manufacturing lead times; tooling, assembly lines, plants and facilities conversion; employment; engineering resources; suppliers' capabilities to meet changes; principal federal and state regulations on environment and safety affecting vehicle design and operation, including new safety standards (e.g., automatic restraints, side and head impact protection, roof crush resistance), Clean Air Act Amendments of 1990, California Air Quality issues, phase-out of chlorofluorocarbons (CFCs); availability and use of alternative fuels; marketability of new vehicles; initial and life-cycle costs of vehicle ownership; competitiveness issues; best-in-the-world vehicles (on the road); prototypes in testing.

It is anticipated that, in the conduct of Tasks 1 and 2, a workshop will be held as described earlier. Proceedings of the workshop will be published promptly on a stand-alone basis.

Task 3

The Committee will prepare estimates by vehicle size class of the fuel economy gains that can be practically achieved in the United States in the next decade. As appropriate, the Committee will condition its estimates in terms of sensitivities expected to selected external factors. Examples of such factors (which may also require assumptions and judgments by the Committee) include the state of the U.S. economy at the end of the decade; world oil prices and availabilities; current product plans of automobile manufacturers; heightened public concerns for safety; and so forth.

The Committee will also prepare estimates, by vehicle size class, of the average incremental first cost per vehicle to the consumer attributable to higher fuel economy (relative to estimates of average life-cycle costs of vehicle ownership and operation), and the incremental annual cost, in aggregate, to the automotive industry in producing higher fuel economy vehicles.

The Committee will not, however, address the formulation of new corporate average fuel economy (CAFE) standards using its estimates of practically achievable fuel economy improvements in new vehicles nor will it, in Phase 1, address other public policy measures to achieve greater fuel economy in new vehicles.

Task 4

The Committee will prepare a Phase 1 report setting forth its findings, the rationale therefor, and the description of the barriers identified in this Phase. A manuscript of this report (after it has been subjected to the National Research Council review process) will be delivered by the National Research Council to the National Highway Traffic Safety Administration by June 30, 1991.

PHASE 2

Task 5

As presently envisioned in Phase 2, the Committee will analyze in greater detail the principal barriers to the market introduction and adoption of the most important technologies considered in Phase 1 and present alternative approaches to overcoming these barriers. The Committee will also consider addressing technologies such as electric and hybrid vehicles that were not considered in Phase 1. In any event, a more specific definition of Phase 2 requirements will be made in conjunction with the National Highway Traffic Safety Administration on the basis of information generated in Phase 1.

Task 6

The Committee will prepare a Phase 2 report setting forth its findings and conclusions. A manuscript of this report (after it has been subjected to the National Research Council review process) will be delivered by the National Research Council to the National Highway Traffic Safety Administration by March 31, 1992.

ANTICIPATED RESULTS

The study will result in two reports, one at the end of each phase, and a published proceedings of a workshop, which will be held in Phase 1.

Committee Addenda:

- *In consultation with NHTSA subsequent to the committee's first meeting, May 13-15, 1991, the date for completing Phase 1 of the study was extended beyond June 30, 1991.*

- *The committee held a workshop as part of its second meeting, July 8-12, 1991. The proceedings of the workshop could not, however, be published as originally planned because of time and resource constraints.*

- *The schedule for Phase 2 of the study has not yet been determined.*

APPENDIX B

PROVEN AUTOMOTIVE TECHNOLOGIES: FUEL ECONOMY AND PRICE IMPLICATIONS

This appendix (1) describes how each proven fuel economy technology works and the aspects of vehicle energy use it affects, (2) examines and compares literature estimates of the improvements in fuel economy that may be achievable for each alternative technology compared with a baseline technology, and (3) examines literature estimates of the retail price equivalent (RPE) of using each alternative technology. The appendix then develops the data bases that underlie the technology-penetration, or shopping cart, projections of fuel economy in Chapter 7.

DATA SOURCES FOR THE SHOPPING CART PROJECTIONS

To implement the shopping cart approach, one must have data on the costs, fuel economy contribution, and market penetration for the technologies of interest. All are difficult to acquire. In practice, costs proved to be more difficult for the committee to estimate than potential fuel economy improvement, because the underlying bases for the costs are less well defined and hitherto not well analyzed. Also, information on costs is proprietary in nature so the open literature is very sparse.

The committee obtained data on the market shares of the technologies in MY 1990 from Energy and Environmental Analysis, Inc. (EEA, personal communication, October 2, 1991) and from SRI International (1991). The EEA provided the committee with estimates of the market shares for the various technologies by size class and by import versus domestic manufacture, for passenger cars and light trucks. The SRI report provided estimates for all passenger cars manufactured by members of the Motor Vehicle Manufacturers Association (MVMA), that is, Chrysler, Ford, General Motors, and Honda of America.[1] The committee compared the two sources by computing the sales-weighted average market shares for domestic cars based on EEA's

---

[1]The text of the SRI report might suggest that the cost analysis included inputs from all the manufacturers. Honda informed the committee that it did not contribute cost information to the study.

data and comparing them with the market shares reported for domestic manufacturers in the SRI report. By and large, the estimates are in good agreement. Differences (e.g., market shares of 4-valve engines) seem to arise from the inclusion of Honda's U.S. production in the SRI data and its exclusion from EEA's domestic estimates (Honda's U.S. production is considered imported for corporate average fuel economy [CAFE] purposes).

The percentage improvement in fuel economy that can be ascribed to a given technology continues to be debated among scientists and engineers. While there has been agreement on some technologies, the committee found contention about others. Most of the arguments have to do with the definitions of technologies--the same name is often given to quite different versions of a generic technology in different sources. Some differences have to do with the details of how a technology is implemented. Most engine technologies considered, for example, can be optimized for performance or fuel economy. When optimized for performance, they do not yield as great a fuel economy benefit.

The automotive industry and the U.S. Department of Energy (DOE), together with EEA, a DOE contractor, have spent a considerable amount of time and effort attempting to resolve the debate over fuel economy potential. In meetings over nearly two years, engineers and experts from the domestic manufacturers and DOE have scrutinized definitions, assumptions, and estimation methods. This process produced revisions of several estimates and a narrowing of differences, but not complete agreement. Estimates made by nearly all the major automobile manufacturers have been compiled by Ford Motor Company (1991). Estimates for particular sets of technologies have also been developed by Berger et al. (1990) and by SRI (1991). The committee considered all these sources, which are compiled in Table B-1. It elected to base its shopping cart projections on two sets of estimates--those developed by EEA (1991a) under the sponsorship of DOE and those developed by SRI (1991) under the sponsorship of the MVMA. The SRI estimates, developed to serve as a consensus from the domestic industry, are generally similar, but not identical to the estimates provided by Ford. The EEA and SRI reports are the only sources that provide technology-specific information on both percentage fuel economy improvements and costs. The cost estimates are summarized in Table B-2.

ENGINE TECHNOLOGIES

Under the category of engine technologies in Table B-1 are included those technologies that address the thermodynamic efficiency of combustion, internal engine friction, and pumping losses, as well as energy used by essential engine accessories, such as oil pumps and alternators, and nonessential accessories, such as air-conditioners and power steering.

TABLE B-1 Estimates of Fuel Economy Improvement Potential of Various Technologies (percent)

TECHNOLOGY	BASELINE	EEA	SRI	BSA	FORD	GM	CHRYSLER	TOYOTA	HONDA	NISSAN	MITSUBISHI
ENGINE TECHNOLOGIES											
GENERAL											
Roller cam followers	Flat followers	2.0	1.7	0.3	3.0	1.5	2.4	0.8	1.0	1.4	1.3
Friction reduction, -10%	Base 1987	2.0	2.0		2.0	1.0	0.5	0.8	1.0	1.4	
Accessory improvement	Conventional	0.5	0.7		0.7	0.0	1.4	0.5		0.2	0.8
Deceleration fuel restriction	None	1.0	1.0		1.0						
Compression ratio, +.5	9:1 (EEA 4-V only)	[a]	2.0		1.5	1.0		1.3			1.0
FUEL SYSTEMS											
Throttle-body fuel injection	Carburetor	3.0	2.6	3.0	3.0	2.5	3.4	0.8	1.0	3.3	
Multipoint fuel injection	Carburetor	5.0 [b]	4.6	3.1	6.0	4.0	4.9	2.5	3.5	4.3	
VALVE TRAIN											
Overhead camshaft	Overhead valve	3.0	2.5	1.2	3.5	1.5	2.0		0.8	2.0	
4 valves per cylinder	2 valves	5.0	3.0	2.1	3.5	3.0	3.5	4.5	2.0	3.4	
Variable valve timing	Fixed timing	6.0	2.6		3.0	2.0	1.5	2.0	2.5 [c]	2.7	
REDUCED NUMBER OF CYLINDERS											
4-cylinder	6-cylinder	3.0	0.0	1.2	-3.0	0.0	0.0	0.0	0.0	0.0	
6-cylinder	8-cylinder	3.0	1.0	-0.9	0.0	0.0	0.0	0.0	0.0	0.0	
TRANSMISSION TECHNOLOGIES											
Torque converter lock-up	Open converter	3.0	2.0	2.8	2.0	3.0	3.0	2.5	3.0	3.2	
Electric transmission control	Hydraulic	0.5	0.5	0.5	0.5	0.0	0.5	0.5	0.5	0.6	
4-speed Automatic	3-speed auto	4.5	2.8	2.9	3.0	4.0	2.0	2.3	1.8	3.0	
5-speed Automatic	3-speed auto	7.0	3.3		5.0	4.5	3.0	3.5	3.3	4.0	
Continuously variable transmission	3-speed auto	8.0	4.8		5.5	4.5	3.0		3.8	5.5	
5-speed Manual [d]	3-speed auto	8.0	4.8	0.0	5.5	0.0	0.0	0.0	0.0	0.0	
ROLLING RESISTANCE, AERODYNAMICS, AND WEIGHT											
Front wheel drive	Rear wheel drive	10.0	0.5	0.8	1.0	0.0			1.1	3.0	
Aerodynamics	Base	2.3	2.4	2.7	2.0	3.1	2.0	2.0	1.5	1.2	1.7
Weight reduction, -10%	Base	6.6	5.0	9.1	5.5	8.0	5.0	5.5	5.0	6.0	
Electric power steering	Conventional	1.0	1.4		1.5	0.5	1.0	1.0	1.0	1.0	
Advanced tires, -10%	Base	1.0	1.0	0.6	1.0	0.5	0.5			1.0	1.0
Advanced lubricants	Conventional	0.5	0.3		0.2			0.5	0.5		

[a] Fuel economy benefit for EEA incorporated into 4-valve engine.
[b] Apportioned to account for incorporation of limited deceleration fuel restriction in multipoint fuel injection.
[c] A savings as large as 12.5 percent can be inferred from discussion in Chapter 2 or Appendix C.
[d] Fuel economy benefit assumed same as that of CVT over 3-speed automatic transmission.
Source: Committee adaptation of summary of presentations to the committee, July 1991, prepared by A. Gilmour (Ford, 1991). Baseline technologies are arbitrary and have been changed from some original sources to put all estimates on a comparable basis.

TABLE B-2 Costs of Fuel Economy Improvement Technologies

		Data Source and Engine Type					
		EEA (1988 $)			SRI (1990 $)		
TECHNOLOGY	BASELINE	4 Cyl	6 Cyl	8 Cyl	4 Cyl	6 Cyl	8 Cyl
ENGINE TECHNOLOGIES							
GENERAL							
Roller cam followers	Flat followers	16	24	32	65	65	65
Friction reduction, -10%	Base 1987	30	40	50	60	60	60
Accessory improvement	Conventional	12	12	12	200	200	200
Deceleration fuel restriction	None				5	5	5
Compression ratio, +.5	9:1 (EEA 4-V only)				1	1	1
FUEL SYSTEMS							
Throttle-body fuel injection	Carburetor	42	70	70	65	65	65
Multipoint fuel injection	Carburetor	90	134	150	215	215	215
VALVE TRAIN							
Overhead camshaft	Overhead valve	110	160	200	400	400	400
4 valves per cylinder	2 valves	140	180	225	400	400	400
Variable valve timing	Fixed timing	140	200	267	100	100	100
REDUCED NUMBER OF CYLINDERS [a]							
4-cylinder	6-cylinder	0	(300)	(550)	0	(300)	(550)
6-cylinder	8-cylinder	300	0	(250)	300	0	(250)
TRANSMISSION TECHNOLOGIES							
Torque converter lock-up	Open converter	50	50	50	56	56	56
Electric transmission control	Hydraulic	24	24	24	122	122	122
4-speed Automatic	3-speed auto	225	225	225	230	230	230
5-speed Automatic	3-speed auto	325	325	325	530	530	530
Continuously variable transmission	3-speed auto	325	325	325	640	640	640
5-speed Manual [d]	3-speed auto						
ROLLING RESISTANCE, AERODYNAMICS, AND WEIGHT							
Front wheel drive	Rear wheel drive	240	240	240	26	26	26
Aerodynamics	Base	40	40	40	60	60	60
Weight reduction, -10%	Base	--	varies [b]	--	470	470	470
Electric power steering [c]	Conventional	45	45	45	61	61	61
Advanced tires, -10%	Base	18	18	18	20	20	20
Advanced lubricants	Conventional	2	3	3	3	3	3

[a] Reduced number of cylinders keeping engine displacement constant. Numbers in EEA columns are based on SRI.
[b] Based on cost of $0.50 per pound saved (EEA, 1991a) multiplied by 10 percent of average weight of all cars in the size class.
[c] Committee estimate based on price of electric power steering for Honda Civic in Japan.
Source: Committee estimates based on adaptation of data from EEA (1991b), SRI (1991), and other sources.

General

This subcategory of engine technologies includes those specifically addressing friction reduction and thermodynamic efficiency, as well as certain ones that do not fit under the other subcategories--fuel systems, valve trains, and number of cylinders.

Roller Cam Followers

In conventional engines, intake and exhaust valves are operated by a camshaft whose lobes are in sliding contact with a cam follower. This is a large source of friction in a conventional engine, accounting for up to one-fourth of all engine friction (Ledbetter and Ross, 1990). Roller cam followers incorporate hardened steel roller bearings that reduce this source of friction. They are estimated to increase fuel economy by about 2 percent. Domestic manufacturers tend to give higher estimates than foreign manufacturers, as shown in Table B-1, and they currently make much greater use of roller cam followers, which are already in widespread use in car and light-truck engines of all sizes.

EEA (1991b) estimates that the RPE of roller cams is $4 per cylinder, or $16 for a 4-cylinder engine to $32 for an 8-cylinder. SRI (1991) reports a much higher RPE, $65, as an average for all cars.[2]

Friction Reduction

About 20 percent of engine power is lost to friction (Office of Technology Assessment [OTA], 1991). The primary sources of friction at moderate engine speeds, in order of importance, are the pistons and rings, valve train, crankshaft, and oil pump (EEA, 1991a). Engine friction has been gradually reduced over several decades. According to SRI, redesign of pistons and rings and modification of bearings throughout the engine could produce an overall 10 percent reduction in engine friction, yielding a fuel economy gain of 1.5 to 2.0 percent. EEA and Ledbetter and Ross (1990) concur with the high end of this range (2.0 percent) for the fuel economy effect of low-tension piston rings, closer machining tolerances for pistons, cylinders and bearing surfaces, and use of lightweight pistons.[3] The latter sources point out that the use of lightweight valves and ceramic pistons, titanium valve springs, lightweight composite connecting rods, and two rather than three piston rings, together with oil-pump and crankshaft modifications, could reduce engine friction by another 10 percent, for another 2 percent fuel economy benefit. SRI considers lightweight valve trains separately and estimates a fuel economy improvement of 0.5 percent for that change alone.

Overall, then, a fuel economy improvement of 2 percent for each 10 percent reduction in engine friction, up to a maximum friction reduction of 20 percent, seems to be a reasonable estimate. Although the amount of friction reduction achievable and its impact on fuel economy may vary by engine, there are no inherent limitations on the use of friction-reducing technology in the engine.

---

[2]In this appendix all EEA cost estimates are quoted in 1988 dollars and the SRI estimates are quoted in 1990 dollars.

[3]Advanced synthetic lubricants give small additional reductions in friction. However, they are expensive and, to date, the Environmental Protection Agency (EPA) has not permitted their use in fuel economy tests for CAFE purposes because it cannot be guaranteed, owing to their cost, that they will be used in the field.

The RPE estimates for a 10 percent reduction in internal engine friction are about $50 per car. SRI (1991) puts the RPE at $60, and EEA (1991b) puts the RPE at $30 for a 4-cylinder, $40 for a 6-cylinder, and $50 for an 8-cylinder engine.

Accessory Improvements

Accessories either perform essential engine-supporting functions (e.g., the water pump, oil pump, cooling fan, and alternator) or provide optional services for the driver and occupants (the power-steering pump and air-conditioning compressor). They can account for perhaps 15 percent of vehicle energy requirements (EEA, 1991a). Accessories typically require about the same amount of energy regardless of vehicle size, so they have a somewhat greater proportional impact on the fuel economy of smaller cars. The energy requirements of accessories do not typically increase in direct proportion to engine speed, yet traditional accessory drive mechanisms are geared so that their speed does increase with the engine speed, which results in a poor match between energy inputs and requirements.

Fuel economy can be improved by increasing the efficiency of the accessory system or by better matching its operation to requirements. A great deal of improvement has already been achieved in this area over the past decade. For example, before 1980, most cooling fans were driven by a drive belt operating from the crankshaft. The faster the engine speed, the faster the fan turned. However, at highway speeds the fan is not usually needed, so the energy used to run the fan was wasted. Today, front-wheel drive vehicles are equipped with thermostatically operated electric fans that turn on only when needed.

More generally, accessories driven by a single-speed drive use excessive energy at high engine speeds (SRI, 1991). Variable-speed drives can reduce this waste, but so far the cost and complexity of variable-speed drive systems have not been justified by the 0.5 to 1.0 percent efficiency improvement they can achieve (EEA, 1991a; SRI, 1991). EEA asserts, however, that incremental improvements in drive systems, optimization of fan and pump blade shapes, and reduced heat rejection from the engine can combine to raise fuel economy.

Estimates of the costs of accessory improvements differ, depending on which specific improvements are included. EEA (1991b) estimates an RPE of $12 for an 0.5 percent improvement, excluding use of variable-speed drives and electric power steering. SRI (1991) estimates that the RPE of two-speed accessory drive will be $200. The high and uncertain costs of these technologies support the committee's view that variable-speed drives are not proven technology.

Deceleration Fuel Restriction

Since the momentum of the vehicle actually drives the engine during deceleration, it is possible to restrict the fuel input sharply with no effect on operation. In the extreme, shutting off all fuel flow would require restarting the engine to restore power. This is the version considered by SRI (1991). EEA (1991a) combines a partial reduction in fuel flow that does not shut off the engine with multipoint fuel injection. Both SRI and EEA conclude that fuel restriction during deceleration can increase fuel economy by about 1 percent. There are no technical limits on the applicability of this technology. SRI (1991) puts its RPE at $5; EEA (1991b) bundles it with multipoint fuel-injection technology, discussed below.

Compression Ratio Increase

All else remaining equal, an engine with a higher compression ratio converts a greater proportion of fuel energy into useful work and a lesser proportion to waste heat; that is, the engine has a higher thermal efficiency. Small changes in engine design to ameliorate the increased tendency to knock (modification of the cylinder heads, electronic engine controls, and addition of knock sensors) can lead to a 5 to 6 percent increase in compression ratio (typically from about 9.0:1 to 9.5:1) with an accompanying 1.3 to 2.0 percent increase in fuel economy (SRI, 1991) and without requiring use of higher octane fuel. Higher compression ratios are generally associated with greater production of oxides of nitrogen (NO_x). Such modifications could be made to essentially all gasoline-powered vehicles, although they are most suitable for vehicles that do not already use high compression ratios. The RPE of a compression-ratio increase is estimated by SRI to be very small--on the order of $1--assuming no additional hardware or controls are needed.

Fuel Systems (Throttle-Body and Multipoint Fuel Injection)

In 1975, 95 percent of all passenger cars and 99.9 percent of all light trucks sold in the United States used carburetors. By the 1991 model year the situation was completely reversed: 99.7 percent of all cars and 98.1 percent of all light trucks were equipped with fuel injection (Heavenrich et al., 1991).

Fuel injection has several advantages over carburetion. Because the flow restriction of the carburetor is eliminated and there is no need to preheat the air/fuel mixture, torque and maximum horsepower are increased (Newton et al., 1989). In addition, fuel injection controlled by modern computer-based electronics can better match fuel supply to engine operating conditions, thereby improving drivability, emissions control, and fuel economy. Compared with carburetion, fuel injection leads to a modest reduction in pumping losses in getting air/fuel mixture into the combustion chamber and to a slight reduction in the relative importance of engine friction because power per unit of displacement is increased. Fuel injection is a clear example of a technology that offers significant benefits beyond increased fuel economy.

There are two general types of fuel-injection systems. Throttle-body (or single-point) fuel injection uses one or two injectors to inject fuel upstream of the throttle valve at essentially the same place as from a carburetor. Multipoint fuel injection (MFI) locates an injector immediately upstream of each inlet valve, which enables better control of the air/fuel mixture to each cylinder. Throttle-body injection (TBI) is simpler and cheaper than multipoint injection.

EEA (1991a) estimates the combined fuel economy improvement of TBI at 3 percent, and SRI (1991) suggests a 2.6 percent improvement. Three percent seems a good consensus estimate, as suggested by Table B-1.

According to EEA (1991a), MFI produces an additional 1.2 to 1.5 percent improvement in fuel economy over TBI. SRI (1991) gives a slightly higher estimate of 1.5 to 2.0 percent. MFI allows greater control of fuel flow during deceleration, and it is required for effective deceleration fuel shutoff. EEA includes this effect under MFI, and SRI reports it separately under "Deceleration Fuel Off." EEA points out that MFI also allows the use of a tuned intake manifold to optimize airflow and that the unheated charge is denser than a preheated charge would be, which increases volumetric efficiency. As a result, EEA points out that an optimized drivetrain would adjust the rear-axle ratio for the slight increase in torque with MFI, which would produce another gain in fuel economy of approximately 0.5 percent. In sum, SRI and most manufacturers suggest a 1.5 to 2 percent gain for MFI over TBI, whereas EEA indicates a 2 percent gain, with an additional 1.0 percent available through deceleration fuel restriction. Thus, the total improvement over a carbureted system is about 5 percent without deceleration fuel shutoff, and about 6 percent with.

Both TBI and MFI systems are applicable to all passenger cars and light trucks with spark-ignition engine. Because of the multiple advantages of the MFI system, it is likely to replace TBI systems by 1995 or shortly after (SRI, 1991).

Cost estimates for TBI systems are in close agreement. EEA (1991b) estimates the RPE at $42 for one-injector and $70 for two-injector systems, while SRI (1991) indicates $65 without specifying the number of injectors (presumably one). These sources do not agree, however, on the RPE of MFI. EEA's estimates range from $48 for a 4-cylinder engine to $80 for 8-cylinders, and SRI reports $215 without specifying engine size.

Valve Train Technologies

Valve train improvements can improve engine efficiency in three areas: (1) pumping losses, (2) engine friction, and (3) thermodynamic efficiency.

Overhead Camshaft

Locating the camshaft above the cylinder heads to operate the valves directly allows several improvements in engine design. Inertial forces in the valve train are reduced because some parts of the traditional arrangement are eliminated. Having fewer moving parts also reduces friction and improves high-speed operation by allowing the valves to be opened and closed more rapidly. Similarly, the total valve-opening time can be reduced, which improves low-speed torque and fuel economy. Finally, some overhead cam (OHC) designs allow increased flexibility in valve location and, therefore, improved shape of the combustion chambers.

All of the above factors allow greater power output for a given engine size. Overhead-valve engines of older design achieved power outputs of about 40 BHP/liter (brake horsepower per liter of displacement), and modern versions achieve 45 BHP/liter. In comparison, modern OHC engines produce 50 to 55 BHP/liter.

At equal power output, OHC engines achieve better fuel economy than overhead-valve engines, although the exact difference depends strongly on the related changes and design choices that are made. SRI (1991) suggests gains in the range of 1.1 to 2.5 percent without design changes; the General Motors, Ford, and Chrysler estimates are 1.5, 2.0, and 3.5 percent, respectively (see Table B-1). EEA (1991a) reports a 1.0 percent efficiency gain at constant displacement, a 3.0 to 3.8 percent efficiency gain due to a reduction in displacement to achieve constant peak power, and a 1.1 to 1.3 percent *loss* in efficiency due to a change in axle ratio to compensate for changes in the shape of the torque curve--all of which results in an overall gain of 2.9 to 3.5 percent in fuel economy. There are no limits to the applicability of OHC engines.

The cost penalty of an OHC engine depends primarily on the complexity of the camshaft drive system. There is little inherent reason for an OHC engine to cost more since it has fewer moving parts and does not require any exotic technologies. EEA (1991b) suggests an RPE of $110 for a 4-cylinder engine, $160 for a 6-cylinder, and $200 for 8-cylinders. SRI (1991) reports an average RPE of $400 per engine. The committee believes that both of these estimates may be too high, especially for 4-cylinder engines.

Four Valves per Cylinder

Conventional engines use two valves per cylinder, one each for intake and exhaust. As engine speed increases, the aerodynamic resistance to pumping air in and exhaust out of the cylinder increases. By doubling the number of intake and exhaust valves per cylinder, pumping losses are reduced and useful power output is increased, especially at high engine speeds. Still greater improvements can be achieved by using variable valve timing and lift control to take advantage of the 4-valve configuration (see below). In many cases, the inlet passage to each valve is controlled separately, which further improves the operation of the engine over wide ranges of speed and load.

In addition to enhancing the flow of gases in the engine, the 4-valve design also allows the spark plug to be positioned closer to the center of the combustion chamber, which decreases the distance the flame must travel to complete combustion. In addition, using two streams of incoming gas can help to achieve more complete mixing of air and fuel, further increasing combustion efficiency (Newton et al., 1989). The Honda VTEC-E engine uses this central placement to create optimal conditions for "lean" combustion (see below). Central placement of the spark plug not only promotes more rapid combustion, but also allows the ignition timing to be retarded, thereby decreasing the dwell time of hot gases in the combustion chamber and reducing the formation of NO_x (Newton et al., 1989).

Four-valve engines typically produce 10 percent higher torque and 20 percent greater peak horsepower than OHC engines of 2-valve design (EEA, 1991a; Ledbetter and Ross, 1990). SRI suggests a fuel economy gain of 0.8 to 1.3 percent from use of 4-valves, which permits reduction of the engine displacement while maintaining output power. This estimate apparently ignores the opportunities for more rapid combustion due to a more central location of the spark plug. EEA claims that the spark plug location and improved airflow together allow an increase in compression ratio of about 10 percent (typically to 10:1) without increased octane requirements. EEA breaks down the fuel economy changes resulting from replacing a conventional engine with a 4-valve engine as follows:

10 percent decrease in displacement	+3.8%
5 percent axle ratio increase	-1.1%
Increase in thermal efficiency	+2.5%
Increase in valvetrain friction	-0.5%
Reduced pumping losses	+0.5%
NET FUEL ECONOMY CHANGE	+5.2%

SRI and auto manufacturers suggest that the applicability of 4-valve engines will be limited by their relatively low torque at low engine speeds. Poor low-speed torque would make such engines less suitable for large cars and light trucks. It would also limit their acceptance by consumers who value the acceleration and "feel" of vehicles with high torque at low speed. Fortunately, variable valve timing and lift control have the potential to restore the low-end torque of 4-valve engines and make their torque curves resemble those of 2-valve engines of equal power (see below). In its analysis, the committee assumed that the two technologies are used in combination, so it sees no limits on the use of 4-valve technology in the future.

Four-valve systems are more complex than 2-valve systems and, therefore, are significantly more costly. SRI (1991) suggests that the RPE of a 4-valve single overhead cam (SOHC) engine would be $400 more than a 2-valve SOHC and that a 4-valve dual overhead cam (DOHC) engine would be priced at $650 more than a 2-valve SOHC. EEA (1991b) indicates lower prices: $140 for a 4-cylinder, $180 for a 6-cylinder, and $225 for an 8-cylinder engine.

Variable Valve Timing and Lift Control

Conventional engines use fixed valve timing and lift at all engine speeds. At light loads, closing the intake valve earlier would reduce pumping losses (EEA, 1991a). If valves are opened further and for a longer time at higher speeds, the engine can "breathe" more easily, which produces higher horsepower.

A variety of approaches can be taken to variable valve timing. Honda's lean-burn VTEC-E engine is an example of a 4-valve, variable valve control engine optimized for fuel economy.[4] The VTEC-E achieves 15 percent higher torque at 2,000 rpm, and generally higher torque across the range of low rpm, than an equivalent Honda non-VTEC, 4-valve engine (EEA, 1991a; Honda Motor Company, 1991). At

---

[4]The California version of the VTEC-E uses enhanced exhaust gas recirculation rather than excess air. See Appendix C for a description of lean-burn technology.

low rpm, one of the two intake valves in each cylinder remains nearly closed, which creates a swirling motion in the combustion chamber that allows a rich air/fuel mixture to be maintained in a vortex near the centrally located spark plug, while the total air/fuel charge remains very lean. (The nominally closed intake valve remains slightly open to prevent accumulation of liquid fuel in the port.) This feature, which is made possible by variable valve control, allows smooth operation under lean-burn conditions at low rpm. In a presentation to the Technology Subgroup of the committee, at its meeting on September 5-6, 1991 in Detroit, Michigan (see Appendix F), Honda asserted that these features produce a 10 to 15 percent gain in fuel economy and that variable valve control alone without lean-burn operation yields a 7 to 8 percent fuel economy benefit.

Honda's estimate of the VTEC-E's fuel economy apparently does not take account of reoptimization of the drivetrain to take advantage of the higher low-rpm torque that variable valve control makes possible. The difference in fuel economy between the VTEC-E (in the 1992 Civic VX) and a non-VTEC 16-valve engine (in the 1992 Civic DX) is 32.3 percent for the lean-burn version and 22.2 percent for the California version without lean burn. Thus, the lean-burn feature may lead to a 10 percent increase in fuel economy benefit.

The Honda Civic VXs use other fuel-saving features, including a 5 percent reduction in weight compared with the DX, reduced tire rolling resistance, reduced aerodynamic drag (about 3 percent), and changes in axle and gear ratios to take advantage of the VTEC's better torque curve (American Honda Motor Company, 1991; Duleep, 1991). These changes (except the axle and gear changes) taken together may account for about 6 percent of the 32.3 percent and 22.2 percent improvements (about 3 percent for weight reduction, 0.7 percent for the drag reduction, perhaps 0.5 percent each for improved lubricants and reduced accessory loads, and 1 to 2 percent for the improved tires). These estimates suggest a gain of 16 percent for variable valve control and the changes in axle and gear ratios it permits. Other modifications of which the committee is not aware may reduce the benefit attributable to variable valve control and associated drivetrain optimization, but probably not by more than a few percentage points.

Honda's VTEC-E demonstrates that a combination of 4-valve per cylinder technology and valve control can produce a fuel-efficient engine with good torque at low engine speed. As a result, the committee see no limitations, other than those due to cost, on the application of 4-valve engines or variable valve timing and lift control.

Variable valve timing applied to an OHC engine would have an RPE of $100 on average, according to manufacturers' estimates reported by SRI (1991). EEA's (1991b) estimates are considerably higher: $140 for a 4-cylinder, $200 for a 6-cylinder, and $267 for an 8-cylinder engine.

Number of Cylinders

During the past decade and a half, the power output of automotive engines per unit of displacement has increased substantially, which suggests the possibility of continued engine downsizing.[5] At equal displacement and peak horsepower, an engine with fewer cylinders has fewer moving parts and a lower ratio of cylinder surface area to volume. The first factor tends to reduce engine friction, and reducing the surface-to-volume ratio tends to improve the thermal efficiency of the engine, although the larger displacement cylinder has an increased tendency to knock. Also, 4-cylinder engines are typically about 40 to 50 pounds lighter than 6-cylinder ones producing the same power (EEA, 1991a).

The auto manufacturers assert that reducing the number of cylinders raises both idle rpm and the speed at which the engine begins to "lug" (SRI, 1991). In a 4-cylinder engine, there are only two power pulses per revolution so no power is being delivered by a piston to the crankshaft about one-sixth of the time (Newton et al., 1989). The low frequency and high amplitude of these power pulses significantly increase the vibration levels of a 4- compared with a 6-cylinder engine and to a lesser extent, of a 6- compared with an 8-cylinder engine. Idle speed can be increased to overcome vibration at idle, but that consumes additional fuel, thus reducing fuel economy. To ensure smooth operation in low-speed driving, gear ratios and lockup speeds must be changed in going from a 6- to a 4-cylinder engine, which generally results in lower fuel economy under otherwise similar conditions.

There is no argument that engines with fewer cylinders experience the above problems, but there is considerable disagreement about whether they offer net fuel economy benefits after modification to produce acceptable vibration. EEA (1991a) estimates that replacing a V-6 engine with an in-line 4-cylinder (I-4) engine of equivalent displacement could reduce friction by 15 percent, resulting in a 3 percent fuel economy benefit. EEA also suggests a 1.6 to 3.0 percent fuel economy gain in moving from a V-8 to a V-6. SRI (1991) reports a 1 percent gain for the switch from V-8 to V-6 and no gain for changing from a V-6 to an I-4, because of the other design changes that are required to hold consumer satisfaction constant. Ford (1991) indicates that the changes necessary to keep consumers equally satisfied in moving from 6 to 4 cylinders would actually produce a fuel economy *loss* of 3.0 percent.

Whether reducing the number of cylinders is a valid fuel economy option is important for two other reasons. First, engines with fewer cylinders are cheaper to make. SRI suggests a savings of $250 in going from a V-8 to a V-6 engine, and a savings of $300 in converting from a V-6 to a 4-cylinder. Second, several other fuel economy technologies described below also increase an engine's output per unit of displacement, which allows a reduction in engine size for the same power. Reducing displacement generally reduces vibration, which, in turn, allows a reduction in the number of cylinders and produces synergistic fuel economy and cost benefits.

The committee generally assumed that it is practical to decrease the cylinder count by two in future vehicles. That is, six cylinders can be replaced by four, and eight

---

[5]The horsepower produced per cubic inch of engine displacement for 4-cylinder automotive engines has increased from 0.716 in 1975 to 0.925 in 1991, for 6-cylinder engines from 0.485 to 0.786, and for 8-cylinder engines from 0.446 to 0.643 (Heavenrich et al., 1991).

cylinders by six. The committee further assumes that (1) cylinder count is not reduced for existing 4-cylinder vehicles; (2) changes in cylinder count are in combination with other technological changes that increase peak horsepower per unit of displacement (overhead cam, four valves per cylinder) and preserve low-rpm torque (variable valve timing and lift control); and (3) improvements in engine mountings and other areas can reduce the impact on vibration and noise of using engines with fewer cylinders, but some increases in noise, vibration, and harshness are probably unavoidable. The true consumer costs of such changes, therefore, are understated by their RPEs.

Transmission Technologies

Automotive transmissions are a means for varying the ratio of engine speed to vehicle speed, thus allowing the engine to operate somewhere in the optimum speed range and the automobile to operate over a wider range of speeds than would be feasible if the engine were connected directly to the drive wheels.

Transmissions affect fuel economy in two fundamental ways. First, energy is lost to friction within the transmission itself. Second, the wider the ratio range of the gears in the transmission and the more carefully controlled the transmission shift point, the more the engine can be kept in its most fuel-efficient operating regime over a wide range of operating speeds and loads without sacrificing performance. Both aspects can be manipulated to affect fuel economy.

Torque Converter Lockup

The torque converter of an automatic transmission transfers drive power from the engine to the transmission gears; that is, it serves the same purpose as the clutch in a manual transmission except that it also has torque multiplication capabilities. Both slippage and torque multiplication are present when the vehicle is starting from stop or changing gears to allow the synchronization of engine and gears. However, slippage during cruising wastes energy (Ledbetter and Ross, 1990). A torque converter with lockup eliminates slippage when the vehicle is cruising, which makes the converter 100 percent efficient under these conditions. Lockup may be applied to the top gear only, or to lower gears as well.

If engine and transmission speeds are not perfectly matched when the converter locks, a shock is transmitted to the drivetrain that the driver may be able to feel. In addition, the transmission of engine vibration to passengers increases. This has apparently caused some consumer dissatisfaction in the past, particularly in small 4-cylinder vehicles. Improved transmission controls and reduced vibration in 4-cylinder engines should ameliorate these problems and thereby make the lockup feature broadly applicable to cars and light trucks.

There is general agreement that the lockup torque converter increases fuel economy by about 2 to 3 percent. EEA (1991b) and SRI (1991) agree that it adds just over $50 to the price of a vehicle.

Electronic Transmission Controls

Control of automatic transmissions, which is conventionally executed hydraulically, can be improved by using more precise electronic control of gear shifting, with the result that the transmission will operate in the optimum gear a greater proportion of the time. Estimates of the fuel economy benefits of electronic transmission controls in the early 1980s suggested benefits on the order of 3 to 5 percent (EEA, 1991a). Since then, hydraulic systems have been optimized to produce maximum fuel economy over the EPA test cycle. As a result, SRI and EEA agree on a potential additional increase in fuel economy of only 0.5 percent. Others contend that a 1.5 percent increase is possible, perhaps by sacrificing some smoothness of operation for optimum shifting (Ledbetter and Ross, 1990). This technology is widely applicable to cars and trucks. Price estimates differ by a factor of five. EEA (1991b) suggests $24 and SRI (1991) $122.

4-Speed and 5-Speed Automatic Transmissions

In comparison with standard 3-speed automatic transmissions, the additional gear ratios provided by 4- and 5-speed automatics allow the engine to operate closer to its most fuel-efficient regime more of the time. EEA (1991b) claims a fuel economy benefit of 4.5 percent for 4-speed versus 3-speed automatic transmissions. Estimates provided by Chrysler, Ford, and General Motors are 2, 3, and 4 percent, respectively; and SRI suggests that 2.8 percent is achievable (see Table B-1).

For 5-speed versus 4-speed automatic transmissions, SRI (1991) estimates fuel savings of 0.5 percent and EEA (1991a) estimates 2.5 percent. EEA cites a published study by Nissan and unpublished results of tests by Mercedes Benz and Ford that suggest improvements in the range of 2 to 3 percent for 5-speed versus 4-speed transmissions.

Adding gear ratios tends to increase the size, weight, and cost of the transmission, however. Accommodating a 5-speed automatic transmission in minicompact and subcompact cars would be very difficult. Thus, in this analysis, the committee limited minicompact and subcompact cars and light trucks to 4-speed automatics (or continuously variable transmissions, see below), and projected only limited use of 5-speed automatics in the compact class.

As noted, the increased complexity of automatic transmissions with additional gear ratios adds to their cost. SRI (1991) suggests a price increase of $230 in moving from a 3-speed to a 4-speed automatic, and an additional $300 to add a fifth gear. EEA (1991b) is in close agreement about the incremental cost of a 4-speed transmission ($225), but it suggests that a fifth gear would add only $100 to the price.

Continuously Variable Transmissions

The continuously variable transmission (CVT) is based on an entirely different mechanism for connecting the engine and drive wheels in variable ratios. Instead of a set of intermeshing gears of different diameters, one design now in production uses a continuous, flexible drive belt that engages two variable-diameter pulleys, one connected to the engine and one to the output. Sliding the sheaves of the pulleys together or apart to change their diameters changes the ratio of the rpm of the engine to the rpm of the drive wheels.

Within limits, the CVT offers an infinite number of gear ratios. In addition to allowing operation of the engine at its most efficient point regardless of changes in load and vehicle speed, the jerk-free shifting of gear ratios and absence of shock loading on the drivetrain hold out the possibility of reduced wear and a smoother ride (Newton et al., 1989).

Although there are several CVT designs, only Subaru offers one for sale in the United States. Major obstacles to widespread use of CVTs include the difficulty of control and the inability of existing designs to transmit high torque levels. Microprocessors and electronic controls promise to solve control problems (and also open up the possibility of regenerative breaking; Newton et al., 1989) and increase the fuel-savings potential of CVTs (SRI, 1991).

Newton et al. (1989) suggest that an optimized CVT could increase the fuel efficiency of vehicles engaged in stop-start operation by as much as 22 percent. Estimates for gains using proven technology in passenger cars are much lower. EEA (1991a) suggests that current CVTs can do no better than 5-speed manual transmissions; that is, about a 3.5 percent gain over 4-speed automatics (thus, 8 percent over a 3-speed automatic). Ford and Nissan data suggest 2.5 percent is possible, and General Motors and Chrysler estimate 0.5 and 1.0 percent, respectively, in comparison with 4-speed automatics. SRI (1991) reports that 1 to 2 percent is now possible and that gains in electronic controls are likely to increase the benefit.

EEA (1991b) estimates a price increase for a CVT of $100 per car above a 4-speed automatic and $325 above a 3-speed, and SRI (1991) estimates that a CVT would be priced at $410 more than a 4-speed automatic with torque converter lock-up, which implies an increase of over $600 over a conventional 3-speed automatic.

5-speed Manual Transmission

Essentially all 4-speed manual transmissions have already been replaced by 5-speed manual transmissions. The theoretical benefit of a greater number of gears is the same as for the 5- versus 4-speed automatic transmission. The manual transmission has the additional advantage over an automatic of lower friction and thus greater efficiency. About a 1 percent fuel economy gain over a 5-speed automatic (or 8 percent over a 3-speed automatic without lock-up) is reasonable. The committee does not have price estimates for 5-speed manual transmissions, although the price is surely less than for 3-speed automatics, perhaps $150 less. However, due to limits on consumer acceptance of manual transmissions, the committee does not foresee major shifts from automatic to manual transmissions in the future.

ROLLING RESISTANCE, AERODYNAMICS, WEIGHT, PERFORMANCE

All of the above technologies are related primarily to the efficiency with which the drivetrain converts energy in the fuel into useful work. Fuel economy can also be increased by decreasing the amount of work necessary to propel the vehicle that is, by reducing energy needed to overcome inertia (weight), aerodynamic drag, and rolling resistance. A variety of technologies help in this regard.

Front-Wheel Drive

Compared with traditional rear-wheel drive, front-wheel drive (FWD) connects the engine to the drive wheels through shorter drive connections, requires a more complex front axle and steering system, moves more of the vehicle weight to the front wheels, and facilitates more efficient use of the interior space of the vehicle.

Despite the very widespread use of FWD and the large amount of accumulated experience, the literature offers quite different estimates of its impact on fuel economy and its cost. The differences seem to arise from different definitions of what is included in this technology. The primary benefit of FWD is that it makes it possible to reduce vehicle weight while preserving interior volume. FWD incorporates the driveshaft, rear axle, and differential in a single unit, which saves about 100 pounds. The FWD transaxle is typically a little more efficient than a rear-wheel-drive driveshaft and differential. SRI (1991) suggests an efficiency gain of 0.3 to 0.5 percent, and EEA (1991a) suggests 1.5 percent.

The majority of the fuel economy benefit of FWD comes from redesigning the vehicle to reduce exterior dimensions, which is facilitated by the transverse engine mounting and the absence of a driveshaft tunnel. During the late 1970s and early 1980s, vehicles were extensively downsized through repackaging associated with FWD and through conversion to unibody construction, without significant reduction in interior volume. A nearly complete transformation from rear- to front-wheel drive was achieved at that time. If one compares the average weight per interior volume for different size classes of vehicles currently being sold, the FWD vehicles have 10 to 19 percent lower ratios of weight to interior volume (EEA, 1991a). Weight reductions of this magnitude would increase fuel economy by 6.6 to 12.5 percent. At the midpoint of this range, the resulting fuel economy improvement for FWD, including efficiency improvement and all weight effects, should therefore be between 10 and 11 percent.

The RPE of FWD drive conversion has been estimated at $240 by EEA (1991b) and at $25 or more by SRI (1991). The SRI estimate does not include the costs of repackaging and major weight reduction, however.

Aerodynamics

Aerodynamic drag is a force opposing the motion of a vehicle that results from the resistance of the ambient air to the movement of the vehicle through it. Quantitatively, drag is proportional to the product of the frontal cross-sectional area of a vehicle, the square of its velocity, and its coefficient of drag (C_D). Thus, energy lost in overcoming aerodynamic drag is related to the C_D, which is a function of the shape of the vehicle and the many details of its surfaces.

Reducing cross-sectional area usually reduces the interior size of a vehicle and is thus of limited value in reducing drag. Driving slower reduces drag force, which is a considerable part of the motivation for the 55-mph speed limit. However, changing actual road speeds is not a consideration in EPA's Federal Test Procedure (FTP), so the committee focused on reducing in the drag coefficient as the only available means of reducing drag.

Complete data on drag coefficients for current automobiles are not available. EEA (1991b) has estimated that the average C_D for 1988 model year vehicles was in the range of 0.37 to 0.38. Numerous makes and models are available with drag coefficients in the range of 0.30 to 0.33, and the best available models have drag coefficients below 0.30. Therefore, 10 and even 20 percent reductions in C_D are entirely feasible. Although drag coefficients for light trucks are not likely to go as low as those of passenger cars, the committee sees no reason that a 10 percent C_D reduction for trucks is not equally feasible.

Over the EPA FTP, fuel economy varies with drag (and hence the C_D) with an elasticity of 0.2. That is, a 10 percent reduction in C_D will produce a 2.3 percent increase in miles per gallon (OTA, 1991), and SRI (1991) cites 2.4 percent.

If reductions in the drag coefficient are timed to coincide with the periodic redesign of vehicles, the extra cost should be small. However, as drag is reduced by more than 10 percent, significant costs must be incurred for such changes as flush windows and improved fit of body parts. Reducing C_D below 0.29 may require using a covered underbody, which would have a significant impact on price (EEA, 1991a). EEA (1991b) estimates $32 for the first 10 percent reduction and $48 for the second. SRI (1991) indicates that the first 10 percent reduction will cost more -- about $60.

Weight Reduction

Lighter vehicles require less energy to overcome inertial forces (acceleration, hill climbing, and turning).[6] Over the EPA test cycle, fuel economy is sensitive to simple weight reduction, with an elasticity of 0.5 (SRI, 1991). That is, a 10 percent weight reduction yields a 5 percent fuel economy increase. Holding performance constant, reduced vehicle weight allows reduced engine power and size, which adds to the fuel economy benefit of direct weight reduction. EEA (1991b) suggests an elasticity of 0.66 for the combined effect (this elasticity is smaller than would be suggested by the independent effects of a 10 percent reduction in weight and a 10 percent reduction in displacement).

---

[6]Strictly speaking, it is the mass of the vehicle that determines its inertia. The committee chose to use the commonly understood concept of weight, which is proportional to mass and numerically equal to it in common units such as pounds.

Materials substitution and downsizing of components could yield a 10 percent weight reduction in the post-1995 time period, according to SRI. EEA cites work by the Department of Transportation indicating that increased use of high-strength, low-alloy steel, plastics, aluminum, and graphite-fiber-reinforced plastics could reduce vehicle weights by up to 30 percent. A weight reduction of 10 percent during the time frame of this study seems quite feasible and should be applicable to all light-duty vehicle types.

The cost of materials substitution will depend on the details of what materials are substituted where and on the associated changes in part and component designs. SRI (1991) provides estimates of RPEs for weight reductions of 1, 5, and 10 percent over 1995 model year vehicles weighing 3,000 pounds. The RPEs respectively are $50, $120, and $470, for roughly 30-, 150-, and 300 pound reductions in weight. (The price estimate for the 5 percent weight reduction does not appear to be consistent with the other two since it implies costs of under $1 per pound of weight removed. This difference could reflect a different strategy.) EEA (1991a) cites price estimates of $0 to $0.20 per pound of weight removed using glass-fiber-reinforced plastic body panels, and $0.40 per pound for use of aluminum. These estimates would put the price of a 10 percent weight reduction at less than $150.

Electric Power Steering

Electric power steering is one of many types of accessories. It is treated separately here because of its apparently higher cost and uncertainty about its stage of development. Commonly used hydraulic power-steering pumps use a significant fraction of engine power, particularly at low speeds. Replacing them with an electric motor can produce fuel economy gains of 1 percent or more. EEA suggests that the size and power requirements of motors for electric power steering may preclude its application in large vehicles (EEA, 1991a). SRI estimates that the RPE of electric power steering will be $60 or more.

Advanced Tires, Rolling Resistance

Rolling resistance arises primarily from the generation and dissipation of heat due to the periodic flexing of the tires as they bear the weight of the vehicle and provide driving, braking, and cornering forces while rotating. During ordinary driving, only a very small part of tire rolling resistance is due to slippage between the tire and the road and to aerodynamic drag. On the EPA urban driving test cycle, tire rolling resistance consumes about one-fourth to one-third of the energy delivered to the wheels (Ledbetter and Ross, 1990; MacCready, 1989). EEA (1991a) estimates that a 10 percent reduction in tire rolling resistance produces a 2 percent fuel economy benefit over the EPA test cycle. There seems to be good agreement between EEA and SRI (1991) that tire rolling resistance can be reduced by about 10 percent over the next decade or so. The result should be a 2 percent gain in fuel economy, but customer-driven trends toward high-performance tire designs may eliminate half of this gain according to EEA.

The price of reducing rolling resistance is expected to be modest, on the order of $20 per car according to EEA (1991b) and SRI (1991). There should be no limits to applicability of improved tires. However, some trade-offs with wet traction are expected.

Reductions in Performance

In recent years, typical passenger-car and light-truck performance levels have risen substantially. This is true whether performance is measured in acceleration time or horsepower-to-weight ratios (Heavenrich et al., 1991). Since 1987, 0 to 60 mph times have dropped 11 percent for passenger cars and light trucks. Horsepower-to-weight ratios rose by as much as 16 percent and 17 percent for various classes of cars and light trucks, respectively, over the same period. Had performance levels remained constant since 1987, the committee estimates that, on the assumption that a 1 percent reduction in engine performance (horsepower/weight ratio) is associated with a 0.38 percent increase in fuel economy (see Chapter 7), fleet-average fuel economy would now be about 2 mpg higher.

Reducing performance by reducing engine size yields a significant fuel economy benefit, but at the direct cost of performance if engine technology is unchanged. A 10 percent reduction in engine size can increase fuel economy by 3 to 4 percent. There should be no direct price increase associated with performance reduction; in fact, it should reduce cost. However, to the extent that consumers prefer higher performance levels, consumer satisfaction would be reduced.

REFERENCES

American Honda Motor Company, Inc. 1991. Statement of American Honda Motor Co., Inc., on Automotive Technologies for Fuel Economy before the Subcommittee on Environment, House Science, Space, and Technology Committee, October 2, 1991. Washington, D.C.

Berger, J.O., M.H. Smith, and R.W. Andrews. 1990. A system for estimating fuel economy potential due to technology improvements. Paper presented at the workshop of the Committee on Fuel Economy of Automobiles and Light Trucks, Irvine, Calif., July 8-12. University of Michigan, Ann Arbor.

Duleep, K.G. 1991. Honda's new Civic VTEC-E model. Memorandum for Oak Ridge National Laboratory and U.S. Department of Energy, August 14. Energy and Environmental Analysis, Inc., Arlington, Va.

Energy and Environmental Analysis, Inc. 1991a. Documentation of Attributes of Technologies to Improve Automotive Fuel Economy. Prepared for Martin Marietta, Energy Systems, Oak Ridge, Tenn. Arlington, Va.

Energy and Environmental Analysis, Inc. 1991b. Fuel economy technology benefits. Presented to the Technology Subgroup, Committee on Fuel Economy of Automobiles and Light Trucks, Detroit, Mich., July 31.

Ford Motor Company. 1991. Technology benefit/methodology. Attachment 1 of letter of August 14, from Alan D. Gilmour to Richard A. Meserve, chairman of the Committee on Fuel Economy of Automobiles and Light Trucks: table entitled Comparison of Optimum Technology Fuel Economy Percent Benefits Provided to NAS.

Heavenrich, R.M., J.D. Murrell, and K.H. Hellman, 1991. *Light-duty Automotive Technology and Fuel Economy Trends Through 1991.* Control Technology and Applications Branch, EPA/AA/CTAB/91-02. Ann Arbor, Mich.: U.S. Environmental Protection Agency.

Honda Motor Company, Ltd. 1991. Fuel economy estimate for NAS panel. Presented to the Technology Subgroup, Committee on Fuel Economy of Automobiles and Light Trucks, Detroit, Mich., July 31.

Ledbetter, M. and M. Ross, 1990. Supply curves of conserved energy for automobiles. *Proceedings of the 25th Intersociety Energy Conservation Engineering Conference.* New York: American Institute of Chemical Engineers.

Newton, K., W. Steeds, and T.K. Garrett. 1989. *The Motor Vehicle.* 11th ed. London: Butterworths.

Office of Technology Assessment (OTA), U.S. Congress. 1991. *Improving Automobile Fuel Economy: New Standards, New Approaches*. Washington, D.C.: U.S. Government Printing Office.

SRI International. 1991. Potential for Improved Fuel Economy in Passenger Cars and Light Trucks. Prepared for the Motor Vehicle Manufacturers Association. Menlo Park, Calif.

APPENDIX C

EMERGING ENGINE TECHNOLOGIES AND CONCEPT AND PROTOTYPE VEHICLES

EMERGING ENGINE TECHNOLOGIES

Lean-Burn Engine

Internal combustion engines burn a mixture of fuel and air; the air is the source of oxygen needed to engage in the chemical reaction with the fuel known as combustion, or burning. In standard engines, the ratio of air to fuel is set at or very near that which ensures that there is sufficient oxygen in the mixture to burn all of the fuel, yet not an excess amount of air.

A lean-burn engine is designed and operated so that some excess air over and above that needed for complete combustion is introduced into the combustion chambers. The term *lean burn* is also sometimes used to describe an engine in which exhaust gases, rather than excess air, are used to dilute the air/fuel mixture. In a lean-burn engine, the air/fuel mixture may be homogeneous (well mixed) or stratified (the fuel is concentrated in only a portion of the mixture). The diesel engine uses stratified lean combustion, and the Honda VTEC-E engine (discussed below) uses a small degree of stratification.

Assuming that the rate and completeness of combustion can be maintained, fuel economy increases with the addition of excess air to the air/fuel mixture (Lichty, 1967). However, wide-open-throttle (WOT) power decreases because not as much fuel is burned. Because of its potential for increased fuel economy, the homogeneous, lean-burn approach was investigated extensively in the 1960s and early 1970s as an alternative emissions-control approach to the three-way catalyst, which requires use of a stoichiometric air/fuel ratio (A/F) of approximately 14.6, when gasoline is the fuel, which yields lower fuel economy. However, at that time, the lean-burn engine could not meet emissions and drivability requirements and its development was discontinued. Its current revival is due to its acknowledged fuel economy advantage, combined with the availability of electronic fuel injection, which makes possible the use of lean-burn conditions in selected portions of the driving cycle.

If the air/fuel mixture is homogeneous and excess air is added beyond that required for complete combustion, production of oxides of nitrogen (NO_x) increases up to a maximum and then begins to decrease. However, if recirculated exhaust gas (usually referred to as exhaust gas recirculation, or EGR), rather than air, is used as the diluent, NO_x continually decreases. The explanation is beyond the scope of this report, but it involves the effect of excess oxygen, flame temperatures, and nonequilibrium effects in the decomposition of NO_x.

Figure C-1 shows the relationship of fuel consumption to NO_x emissions for various combinations of excess air and EGR in the air/fuel mixture. The starting point for the discussion is a stoichiometric air/fuel mixture without EGR (the highest point on the upper curve, where EGR = 0% and A/F = 14.6). As excess air is added (moving to the right from the starting point), NO_x production increases, reaching a maximum at an A/F = 17.0. Further addition of air beyond A/F = 17.0 (moving to the left on the lower curve) leads to a continuous decrease in NO_x. However, when EGR rather than air is added to a mixture (moving to the left from the starting point), NO_x decreases continuously up to 20 percent EGR.

Figure C-1 also illustrates the relationship between fuel consumption and mixture composition, with excess air and with EGR. Again, moving to the right from the starting point (EGR = 0% and A/F = 14.6), adding excess air continuously decreases fuel consumption until an A/F of approximately 20 is reached, after which fuel consumption increases. The explanation for this pattern is also complicated. It involves reduction in the burning rate with increased dilution, reduction in engine pumping losses, and heat transfer. The same but weaker trend--decrease, then a slight increase--is observed for fuel consumption as EGR increases.

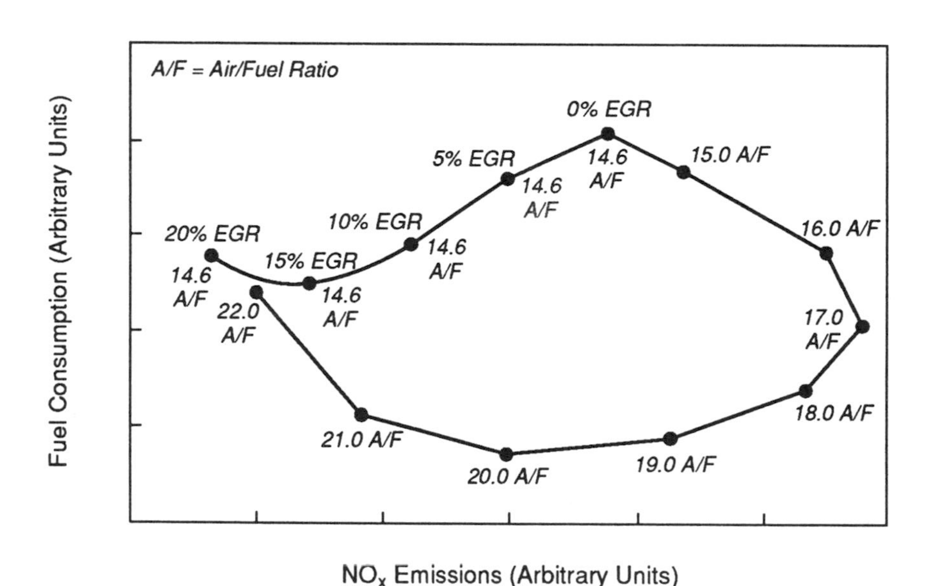

FIGURE C-1 Relationship of fuel consumption to production of oxides of nitrogen (NO_x) in a lean-burn engine at various air to fuel ratios (A/F) with and without exhaust gas recirculation (EGR).

SOURCE: Adapted from Ford Motor Company (1991).

When excess air is used, the conventional three-way catalyst is not effective in reducing NO_x, although it does serve as an oxidizing catalyst for the unburned hydrocarbon (HC) and carbon monoxide (CO) in the lean-burn operating regime if the exhaust temperature is sufficiently high. A three-way catalyst is effective in reducing NO_x when EGR is added to a stoichiometric mixture. However, WOT power decreases as either excess air or EGR is added to the air/fuel mixture in an engine since less fuel is burned. Consequently, stoichiometric or richer (excess fuel) air/fuel ratios without EGR are often used at WOT. And, when either excess air or EGR is added, the burn rate decreases and combustion instability increases, so that HC and CO emissions may become excessive.

The Honda VTEC-E engine gives some clues as to the fuel economy potential of lean-burn engines, although determining how much of its performance is due to the lean-burn feature and how much is due to other changes from the standard engine is difficult. It should be noted that the California VTEC-E version uses EGR to dilute a stoichiometric air/fuel mixture, but the VTEC-E for the other 49 states uses excess air as the diluent. As a consequence, the 49-state version has a 44 percent fuel economy increase while the California version has a 34 percent increase.

The Honda Civic VE with VTEC-E engine differs from the standard Honda Civic not only in the use of lean-burn technology but also in use of variable valve timing; an overall 5 percent weight reduction; improved aerodynamics; reduced rolling resistance through the use of special tires, bearings, and seals; reduced accessory load through the use of a "smart" alternator; reduced engine friction; and a higher rear-axle ratio made possible by the increased torque of the variable-valve-timing system. Because all of these changes can increase fuel economy, it is unclear how much of the fuel economy enhancement of the Civic with VTEC-E over the standard model arises from the engine itself.

Duleep (1991) attributes the 10 percentage point difference in fuel economy improvement between the California and 49-state versions to use of the lean-burn feature in the 49-state model. In Figure C-1, the difference between the minimum fuel consumption using EGR and that using excess air is about 7 to 8 percent. However, since lean-burn conditions are used for only a fraction of the operating regime of engine load and speed, the 10 percent allocation seems generous.

There is a limit to the fuel economy increase obtainable using homogeneous lean burn. Further, HC and CO, as well as NO_x, can be a problem with lean burn. Thus, unless a lean NO_x catalyst is developed or the NO_x emission standard is eased, lean-burn engines will probably be limited to small vehicles or even be excluded from the market completely if emission standards are further tightened.

Diesel Engine

The diesel engine, which is a specific type of lean-burn engine, has been used successfully for decades in large trucks and other heavy applications. Diesels have also been used in a variety of passenger automobiles and light trucks over the past two decades, but their costs, performance, and emissions characteristics keep them off the committee's list of proven technologies. As a consequence of the long experience with them and their promise for the future, they are the subject of continuing development and are considered here as an emerging technology.

Diesels differ significantly from spark-ignition engines. They use much higher compression ratios; they compress air rather than an air/fuel mixture; fuel is injected late in the compression stroke, allowing little time for the introduction, distribution, and mixing of the fuel and air in the combustion chamber; they do not throttle for load control; they employ stratified-charge combustion and lean mixtures; they benefit from turbocharging or supercharging; and they require a fuel quite different from gasoline that will ignite easily when injected into the highly compressed air.

The fuel economy and emissions characteristics of the diesel differ from those of the spark-ignition engine. The diesel's fuel economy is significantly higher, primarily because of its higher compression ratio, its use of turbocharging or supercharging, and its ability to use very lean air/fuel mixtures.[1] Highly supercharged engines, such as those used in heavy-duty trucks, compress the inlet air to a pressure sufficiently high that it is advantageous to cool the air before it enters the cylinder, which increases the density and the mass of air in the cylinder, a further advantage for fuel economy and for emissions control. The low inlet air temperature resulting from cooling reduces the amount of NO_x in the exhaust.

Diesel engines can be divided into two broad categories based on the design of the combustion chamber. The "divided-chamber" diesel has a flow restriction in its combustion chamber, which is divided into two parts. Fuel is introduced into only one section of the chamber, and the combustion-induced flow between the two parts of the chamber provides fuel-air mixing in an extremely short time (a few milliseconds).

The chamber of the "open-chamber" diesel is not divided and depends on the shape of the combustion chamber, the air motion in the chamber induced during intake and compression, and high fuel-injection pressures to accomplish distribution and mixing of the fuel and air. "Direct-injection" diesel engines, which are almost universally used in heavy-duty trucks, are some 12 to 15 percent more efficient than divided-chamber diesel engines. Because they exhibit quieter combustion and greater rpm flexibility, almost all diesels used in passenger cars are of divided-chamber design.

Unless a diesel engine is supercharged, its horsepower per unit of displacement is considerably lower than that of a spark-ignition engine. Since diesel engines generally operate at a lower maximum rpm and the mixing time for the fuel and air is short, it is not possible to utilize all of the oxygen without unacceptable engine emissions. Consequently, for comparable acceleration performance, a diesel engine that is not supercharged must be considerably heavier than a spark-ignition engine.

---

[1] Care must be taken when comparing diesel- and spark-engine fuel economies because the energy content of a gallon of diesel fuel is about 12 to 14 percent greater than for gasoline.

Pollutants emitted from the diesel include solid particulates, as well as the NO_x, CO, and HC emitted from spark-ignition engines. Because of the lean air/fuel ratio and low exhaust temperature, a conventional three-way exhaust catalyst cannot be used. Further, for the diesel there is a trade-off between particulates and NO_x--particulates can be lowered, but at the expense of increased NO_x. Thus, it is doubtful whether the diesel engine can meet future emissions requirements, except possibly Tier I standards. Development of a lean NO_x catalyst would help meet future emissions standards, but such a catalyst would have to meet different requirements from the lean NO_x catalyst needed for lean-burn gasoline engines.

Two-Stroke Engine

Two-stroke engines use an air pump other than the engine's pistons and cylinders to accomplish the four tasks of compression, expansion, intake, and exhaust in two strokes (one revolution of the crankshaft). In contrast, the four-stroke engine requires four strokes (two revolutions) to accomplish the same four tasks.

The most common two-stroke engines are the simple ones used in chain saws and outboard motors. In such engines, openings (called ports) in the cylinder walls serve as valves, and the bottom sides of the pistons, along with the crankcase, serve as a pump to force the fresh air/fuel charge into, and the exhaust charge out of, the cylinder. It is not practical to use a crankcase both as a lubricating oil reservoir and as part of a pump due to entrainment of oil by the air flowing through the crankcase. Consequently, in this configuration, the crankcase is not used as an oil reservoir as it is in four-stroke engines. Instead, a small amount of lubricating oil is added to the fuel or continuously introduced mechanically, and rolling element bearings are used on the crankshaft rather than the sleeve bearings used in four-stroke engines. This configuration gave the two-stroke engine its reputation for light weight, low friction, and high power output. However, the emissions and fuel consumption of this simple configuration are completely unacceptable for modern automotive use.[2]

In the two-stroke engine, exhaust and air or air/fuel intake are accomplished late in the expansion stroke and early in the compression stroke, respectively, by blowing the compressed air or mixture into the cylinder through the intake opening, or port. Ideally, all of the products of combustion (and none of the incoming gases) would be blown out the open exhaust port, leaving primarily fresh gases in the cylinder. However, in practice, a significant portion of the compressed input mixture escapes through the exhaust port and a significant portion of the exhaust gases remains in the cylinder. If the incoming gases contain fresh fuel, the exhaust will contain unburned fuel and the HC content of the exhaust will be high because of the "bypassed" mixture. Offsetting this effect is the fact that the products of combustion that remain in the cylinder with the air/fuel mixture serve as an internal EGR that tends to reduce NO_x emissions.

---

[2]The two-stroke Saab automobiles that reached some acceptance in the United States in the 1960s were incapable of meeting even basic emissions requirements imposed in the late 1960s.

Control of load in the two-stroke engine is achieved by throttling the flow of either the intake or exhaust (which effectively increases internal EGR), by retarding the spark, or by charge stratification as in the diesel engine. All of these load-control techniques have their problems, however. Too much EGR can lead to large cycle-to-cycle variations, spark retard increases fuel consumption, and stratification can increase HC emissions.

The potential benefits from automotive applications of two-stroke technology are reduced engine weight, size, and cost. Each cylinder undergoes a power stroke every revolution, which increases both output power and operating smoothness. If the vehicle is optimized around a lighter, more powerful two-stroke engine, there is a potential for improved fuel economy. Current indications are that the fuel economy of two-stroke engines equals the "best in class" of four-stroke engines for the same vehicle weight.

Current development effort on the two-stroke engine is concentrated on introducing the fuel into the cylinder at a time when no bypassing can occur. This approach would reduce HC emissions and fuel consumption, but it minimizes the time available for mixing the fuel and air.[3]

Significant problems related to mechanical components and exhaust emissions must be overcome before the two-stroke engine can be a serious competitor to the four-stroke engine. A power stroke every revolution is desirable from a smoothness and power standpoint, but it increases the heat load on the pistons since there is less time for them to be cooled by the cool incoming air/fuel mixture. Thus, piston temperature and engine durability have been continuing problems. Combustion stability, especially at idle and at road loads, can also be a problem. NO_x emissions are minimized by the high internal EGR, but HCs and particulate emissions can be a problem. Emissions control, especially to meet California's standards, has not been achieved. Addressing these problems will increase engine weight, cost, and complexity. Thus, the outcome of the current intense developments efforts on two-stroke engines is not clear.

CONCEPT AND PROTOTYPE VEHICLES

A recent concept car of considerable interest is the "efficient personal, experimental" (EP-X) prototype car by Honda. This car, which reportedly gets 100 miles per gallon (mpg) (Levin, 1991), was unveiled at the 1991 Tokyo Motor Show. It seats two passengers in tandem, has an all-aluminum body and a total weight of about 1,400 pounds, and is reportedly powered by a one-liter, lean-burn engine.

---

[3]A variety of configurations are being investigated by different manufacturers (Automotive Engineering, 1991). Toyota has used valves rather than ports to control the timing of cylinder events and a positive-displacement Roots blower for exhaust scavenging and air-fuel supercharging. Subaru uses a screw-type scavenging pump, an electronically controlled fuel-injection system, a spool-type external exhaust control valve, and wet sump lubrication. The 1986 Orbital X 1.2L L3 engine features crankcase scavenging, a spool-type exhaust-port control valve to reduce short circuiting of fresh charge at light loads and idle as well as to improve low-speed torque, and pneumatic fuel injection directly into the cylinder.

Another recent concept car is the Volkswagen Chico (Automotive News, 1991). The Chico is a 1,727-pound hybrid mini hatchback with a top speed of 75 mph, a range of 250 miles, and a clutchless 5-speed manual transmission. According to *Automotive News*, it seats "two adults comfortably, and two small children in the back," meets the pending 1994 side-crash standards, has a latent heat system that stores heat from hot engine coolant for several days for fast warm-up, and "is a hand built concept car [for which] production is not even under discussion." Volkswagen also showed "its diesel-electric Golf featuring both fully automatic drive changeover and engine shutoff/restart and a Jetta electric vehicle using sodium sulfur batteries" (Automotive News, 1991:4).

Bleviss (1988) distinguishes between "High Fuel Economy Production Vehicles" and "High Fuel Economy Prototype Vehicles". Her table for the latter is reproduced here as Table C-1. The distinction between the two categories is important. For example, Amann states, "None of the diesel concept cars have, to my knowledge, demonstrated compliance with upcoming U.S. emission standards. Nor am I aware that any has passed U.S. safety regulations."[4] Presumably all production vehicles, as contrasted with concept cars, have met applicable regulations. Further, as noted by MacCready (1991:2), "Turning a satisfactory new-technology demonstrator into a mass-produced, distributed, and widely applied vehicle takes a long time and a major investment."

Production vehicles, by definition, have met the crucial test of customer acceptance. For the Volvo LCP 2000, Bleviss reports that the "prototype is complete, adaptable to production" (see Table C-1, "Development Status"). However, according to Amann, Volvo has said, "We have found that from a performance point of view, a 5-speed manual gear box is not quite compatible with a high power diesel engine. The alternative would be a 6-7 speed manual gear box which, no doubt, would feel uncomfortable to the average driver ... Test drives with the experimental vehicles have clearly shown that the vibrations and noise of the 3-cylinder diesel engine are unacceptable at engine speeds below 1200-1500."[5] Thus, it is not clear that the Volvo LCP 2000 is in fact "adaptable to production" in its current concept form.

Table C-1 reveals several factors common to high fuel economy prototype vehicles. First, most use diesel engines. The lowest horsepower listed in Table C-1 is 27 and the highest is 88. In contrast, the average horsepower of the 1990 subcompact class was 115. The prototypes are light in weight and most of them make extensive use of aluminum, plastic, and other light materials. The lightest has a curb weight of 1,040 pounds and the heaviest weighs 1,880 pounds. Lovins argues that, with extensive use of composites and plastics, curb weights of 1,000 to 1,400 pounds could be achieved at

---

[4]C.A. Amann, presentation at the workshop of the Committee on Fuel Economy of Automobiles and Light Trucks, Irvine, Calif., July 8-12 (see Appendix F).

[5]Amann, quoting Volvo, Society of Automotive Engineers (SAE) paper 850570; see note 4.

TABLE C-1 High Fuel Economy Prototype Vehicles

Company	Model	Number of Passengers	Aero-dynamic Drag Coeffi-cient	Curb Weight (lb)	Maximum Power (hp)	Fuel Economy (mpg)*	Innovative Features	Development Status
General Motors	TPC (gasoline)	2	.31	1040	38	61 city 74 hwy	Aluminum body and engine	Prototype complete, no production plans
British Leyland	ECV-3 (gasoline)	4-5	.24-.25	1460	72	41 city 52 hwy	High use of aluminum and plastics	Prototype complete
Volkswagen	Auto 2000 (diesel)	4-5	.25	1716	53	63 city 71 hwy	DI with plastic and aluminum parts, fly-wheel stop-start	Prototype complete
Volkswagen	VW-E80 (diesel)	4	.35	1540	51	74 city 99 hwy	Modified DI 3-cyl. Polo, flywheel stop-start, supercharger	Prototype complete
Volvo	LCP 2000 (diesel)	2-4	.25-.28	1555	52, 88	63 city 81 hwy	Hi magnesitum use; 2 DI engines developed, 1 heat insulated	Prototype complete, adaptable to production
Renault	EVE+ (diesel)	4-5	.225	1880	50	63 city 81 hwy	Supercharged DI with stop-start	Prototype complete
Renault	VESTA2 (gasoline)	2-4	.186	1047	27	78 city 107 hwy	High use of light material	Program completed
Peugeot	VERA+ (diesel)	4-5	.22	1740	50	55 city 87 hwy	DI engine, high use of light materials	Ongoing development
Peugeot	ECO 2000 (gasoline)	4	.21	990	28	70 city 77 hwy	2-cylinder engine, high use of light material	Ongoing development
Ford	----- (diesel)	4-5	.40	1875	40	57 city 92 hwy	Di engine	Research
Toyota	AXV (diesel)	4-5	.26	1430-target	56	89 city 110 hwy	Weight is 15% plastic, 6% aluminum, has CVT & DI engine	Ongoing development

Source: Bleviss, THE NEW OIL CRISIS AND FUEL ECONOMY TECHNOLGIES (Quorum Books, New York, an imprint of Greenwood Publishing Group, Inc., 1988), p. 102. Copyright (c) 1988 by Deborah L. Bleviss. Reprinted with permission.

no marginal cost using large, complex assemblies molded as units and snapped together.[6] In contrast, the average curb weight of the subcompact class in the 1990 fleet was 2,520 pounds.

In comparing production with prototype vehicles, it may be instructive to compare the prototypes with the most fuel-efficient 1990 model. The best-in-class, 1990 subcompact car was the General Motors Geo Metro XFI, which weighs 1,750 pounds, uses a 49-hp engine with a manual 5-speed transmission, and achieves a combined city/highway fuel economy of 65.4 mpg. The weight, horsepower, and fuel economy of the Geo are compatible with and in the direction of the values shown in Table C-1. However, it would not be realistic to argue that a car of this size and weight would meet current expectations of the majority of the driving public.

In summary, the constraints imposed on concept vehicles are different from those that must be met by production vehicles. Further, it is not clear that the large weight reductions achieved in prototype vehicles are economically viable, and their effect on safety has not been established. Also, even if acceptable material costs and safety could be achieved, the interior volume of these prototype vehicles is well below that of the current fleet average. This raises questions of customer acceptance, particularly in light of today's low gasoline prices. Such cars, however, might succeed in niche markets, especially in high-density urban areas.

---

[6]A. Lovins, presentation at the workshop of the Committee on Fuel Economy of Automobiles and Light Trucks, Irvine, Calif., July 8-12, 1991.

REFERENCES

Automotive Engineering. 1991. Two-stroke engine technology. 99(7):11-14.

Automotive News. 1991. Chico shows VW answers on clean air. September 9:4,43.

Bleviss, D.L. 1988. *The New Oil Crisis and Fuel Economy Technologies*. Westport, Conn.: Quorum Books.

Duleep, K.G. 1991. Honda's new Civic VTEC-E model. Memorandum for Oak Ridge National Laboratory and the U.S. Department of Energy, August 14. Energy and Environmental Analysis, Inc., Arlington, Va.

Ford Motor Company. 1991. Overview of NAS Technology Subgroup/Ford Meetings. Presented to the Technology Subgroup, Committee on Fuel Economy of Automobiles and Light Trucks, Dearborn, Mich., September 6.

Levin, D.P. 1991. Honda ready to show a car that gets 100 miles a gallon. *New York Times* October 17.

Lichty, L.C. 1967. *Combustion Engine Processes*. New York: McGraw-Hill.

MacCready, P.B. 1991. Further than you might think/electric and hybrid vehicles. Paper presented at Conference on Transportation and Global Climate Changes and Long-Term Options, Asilomar, Calif., August 26.

APPENDIX D

VEHICLE SIZE AND OCCUPANT SAFETY: PRIVATE VERSUS SOCIETAL RISKS IN TWO-CAR COLLISIONS

Numerous empirical studies show that the occupants of a larger car are much safer in a two-car collision than those in smaller cars, all else being equal. It may seem obvious that downsizing (i.e., replacing large cars with smaller cars) would, therefore, make all occupants less safe. This appendix uses a simplified model of occupant fatality risk to show that, under certain conditions, this view may not be correct.

To minimize his or her own fatality risk in multicar collisions, it is in the best interests of the individual to buy the largest, heaviest car available. In so doing, however, the individual imposes an extra risk on the drivers of smaller cars with whom he or she may collide. One person's gain (in reduced risk) causes everyone else's risk to increase. Thus, in two-car crashes, the obvious private safety benefit of larger cars is not an obvious benefit from the societal perspective. From the societal perspective the question is, What is the *net* effect of downsizing? In the committee's view, available research does not provide an entirely adequate answer, in part because available studies do not address the full range of issues from the societal viewpoint, and in part because the net effect depends on precisely how size and weight changes occur.

In principle, downsizing part of the fleet could increase, decrease, or leave total fatalities unchanged in two-car crashes. The total fatality risk from two-car collisions in a fleet consisting entirely of large cars is clearly lower than one consisting entirely of small ones. The question that is more to the point, however, is how the total fatality risk changes as the fleet size mix moves away from the current composition. To examine this question, the committee constructed a simplified model of the dependence of total fleet fatality risk on the size composition of the fleet. The model is based on a simplified fleet mix of small, medium, and large cars and on relative risk factors for two-car collisions that are broadly consistent with those found in the literature, but that are also greatly simplified. To illustrate the relationship of downsizing to societal fatality risks, the committee calculated the hypothetical impacts of several changes in fleet mix on total fatalities in two-car collisions. In the calculations, an assumed 5,000 annual occupant fatalities in two-car collisions was used as the baseline for estimating changes in fatalities.[1] The model takes into account only fatalities in two-car collisions. Omitted are single-car accidents, collisions involving three or more vehicles, collisions between cars and trucks, and pedestrian and cyclist fatalities. No account is taken of nonfatal injuries or property damage. Consequently, these analyses are relevant for about 11 percent of all motor vehicle fatalities.

The relative fatality risks to the occupants of small, medium, and large cars in two-car collisions that were incorporated in the model are shown in the matrix in Table D-1. The occupant of a large car hit by a small car is given an arbitrary relative risk index of 1. The relative risk index is proportional to the probability of an occupant fatality in the "hit" car, given that it is involved in a collision with a "hit-by" car. At the other extreme, the occupant of a small car hit by a large car is assumed to be 36 times as likely to die. Note that occupants in small-car to small-car collisions have a relative risk factor of 9, more than twice that of the occupants in large-car to large-car collisions.

In the model it is assumed that the frequency of collisions between vehicles of different sizes is proportional to the product of the fractions of the vehicle fleet made up of each size vehicle. For example, if small cars make up one-third of the fleet and large cars make up one-fourth of the fleet, the probability of a collision between a small and large car is one-third times one-fourth equals one-twelfth. An index of total fatality risk for the fleet is obtained by multiplying the relative frequencies of crashes of each type by their corresponding relative risk factors and summing all the products. Numbers along the diagonal of the resulting matrix must be multiplied by 2 to account for the occupants of both cars in each crash.

Table D-2 shows the matrix of relative collision frequencies, and Table D-3 shows the matrix of components of the fleet fatality risk index for the base fleet, which is assumed to consist of one-third each of small, medium, and large cars--roughly the distribution of all cars on the road today. The fleet fatality index is computed to be 12.22. The impact of changing the fleet mix is then estimated by assuming that the base size distribution and fleet fatality index of 12.22 corresponds to 5,000 fatalities.

Table D-4 shows the effects on fatality risk and total fatalities of different fleet mixes. For example, in the first variation it is assumed that all large cars are downsized to medium cars and the proportion of small cars remains at one-third. This produces a fleet fatality index of 12.00, yielding a 1.8 percent *decrease* compared with the base case, which corresponds to a reduction of 90 fatalities. In this case, making the fleet smaller, on average, leads to a reduction in total fatalities in two-car collisions. The key to this particular reduction in fatalities is that replacing all large cars with medium ones reduces the risk that the large cars previously imposed on the small and

[1] In 1990 there were 13,406 passenger-car occupant fatalities in two-vehicle collisions (see Chapter 3). However, in most instances the other vehicle was a light truck or heavy truck. The actual number of occupant fatalities in two-car collisions was approximately 4,900.

TABLE D-1 Simplified Illustration of Hypothetical Relative Fatality Risks in Two-Car Collisions

"Hit" Car Size	"Hit By" Car Size Small	Medium	Large
Small	9	18	36
Medium	3	6	12
Large	1	2	4

TABLE D-2 Hypothetical Relative Collision Frequencies, Base Fleet

"Hit" Car Size (Fleet Mix)	"Hit By" Car Size (Fleet Mix) Small (1/3)	Medium (1/3)	Large (1/3)
Small (1/3)	1/9	1/9	1/9
Medium (1/3)	1/9	1/9	1/9
Large (1/3)	1/9	1/9	1/9

TABLE D-3 Base Fleet Fatality Indices

"Hit" Car Size	"Hit By" Car Size Small	Medium	Large
Small	2	2	4
Medium	3/9	12/9	12/9
Large	1/9	2/9	8/9

Total Fatality Index = 12.22

TABLE D-4 Fatality Consequences of Changes in Fleet Size Mix (Hypothetical Illustration Based on Two-Car Collisions Only)

Fleet Name	Fleet Composition			Fleet Fatality Index	Change in Fatalities	
	Small	Medium	Large		Percent	Number[a]
Base Case	1/3	1/3	1/3	12.22	---	---
Small Fleet 1	1/3	2/3	0	12.00	-1.82	-90
Small Fleet 2	2/5	1/2	1/10	12.34	+0.96	+50
Small Fleet 3	1/2	1/2	0	12.75	+4.32	+215
Small Fleet 4	1/4	3/4	0	11.81	-3.35	-170
Small Fleet 5	3/4	1/4	0	14.81	+21.2	+1,060
Small Fleet 6	2/3	2/9	1/9	14.89	+21.8	+1,090
Very Large Fleet	1/9	2/9	2/3	9.70	-20.6	-1,030
All Small	1	0	0	18.00	+47.3	+2,315
All Medium	0	1	0	12.00	-1.82	-90
All Large	0	0	1	8.00	-34.5	-1,725

[a]Based on an assumed 5,000 occupant deaths in two-car collisions.

medium-sized ones by more than the increased risk that is newly borne by the former occupants of large cars who are now in medium-sized ones.

A different, smaller fleet (Small Fleet 2) might feature 40 percent small cars, 50 percent medium, and 10 percent large ones. In this case, the total fleet fatality index increases by about 1 percent, to 12.34, and total fatalities grow by the same percentage, or by 48 deaths.

A number of cases in Table D-4 illustrate that reasonably large changes in fleet mix are associated with changes in fatalities in two-car collisions of plus or minus 5 percent or less (consider Small Fleets 1 through 4). On the other hand, radically smaller fleets yield, according to the model, substantial fatality increases (consider Small Fleets 5 and 6 and the All Small Fleet). Similarly, much larger fleets yield sharp fatality reductions (consider the Very Large and All Large fleets).

This simplified model calculation illustrates two key points in two-car collisions only. First, the changes in societal fatality risks in two-car collisions from downsizing are relatively small compared with the changes in individual private risks, once both winners and losers are considered. Second, the net effect of downsizing is likely to increase total fatality risks in two-car collisions, but the effect *could* be positive, negative, or negligible depending on the precise relative risk relationships and the way in which downsizing occurs.

This simplified analysis has not included collisions with trucks, single-vehicle accidents, or rollovers, in which downsizing is likely to be a liability, nor collisions with cyclists and pedestrians, in which size reduction could be a benefit. Because substantial numbers of fatalities occur in these categories, including them could significantly affect the estimates of the impact of downsizing on total fatalities.

Because of the complexities of the dependence of the total motor vehicle fatality risk on the fleet size mix, the committee urges further examination of these relationships from the societal perspective using real-world data. These analyses should consider all collision types and the full range of crash severity, and they should include a more thorough examination of potential changes in the vehicle fleet mix.

APPENDIX E

SHOPPING CART PROJECTION METHOD: AN ILLUSTRATION FOR SUBCOMPACT CARS

This appendix illustrates the development of the curves of cost versus fuel economy improvement for new vehicles in various size classes used in the shopping cart projection method of Chapter 7.

For each of the technologies considered for improving the fuel economy of new cars in the subcompact size class, Table E-1 shows the committee's estimates of the percentage improvement in fuel economy (column 1) and the associated cost (retail price equivalent, or RPE; column 2) based on the work of Energy and Environmental Analysis, Inc. (EEA, 1990a,b). Columns 3 and 4 of the table show, respectively, EEA's estimates of the current (MY 1990) market penetration of each of the technologies in the subcompact class (EEA, personal communication, October 2, 1991) and the maximum market penetration estimated by the committee. The difference between the two numbers yields the change in market penetration for each technology (column 5).

Column 6 of Table E-1 shows the average percentage improvement in fuel economy arising from increased use of each technology in subcompact cars, calculated as the product of columns 1 and 5. Column 7 shows the average cost of increased use of each technology in subcompact cars, calculated as the product of columns 2 and 5.

Table E-2 displays the technologies from Table E-1, rank ordered by decreasing cost-effectiveness. The effectiveness of each technology is calculated by multiplying the market-share weighted average percentage gain in fuel economy for the technology (column 6, Table E-1) by the base average fuel economy for the subcompact car class of 31.4 mpg. The cost of each technology shown in Table E-2 is carried over from column 7 of Table E-1.

The cumulative cost (RPE) and fuel economy for any ordered subset of the technologies are then developed by summing over the costs and fuel economy increments of the rank-ordered technologies, as shown in the two right-hand columns

Table E-1 Technologies for Improving the Fuel Economy of Subcompact Cars (based on EEA data)

Technology	Fuel Economy Increase [a] (%)	Cost of Technology [a,b] (1988$)	Market Penetration (%) 1990 [c]	Market Penetration (%) Maximum	Change in Market Penetration (%)	Fuel Economy Increase x Change in Market Penetration (%)	Cost x Change in Market Penetration [d] (1990 $)
	(1)	(2)	(3)	(4)	(5)	(6)	(7)
Engine Technologies							
General							
Roller cam followers	2.0	16	29.2	100.0	70.8	1.416	12.46
Friction reduction, -10%	2.0	30	12.3	100.0	87.7	1.754	28.80
Accessory improvement	0.5	12	0.0	100.0	100.0	.500	13.01
Deceleration fuel restriction	1.0	0	58.0	100.0	42.0	.420	0.00
Compression ratio, +.5	0.0	0	0.0	100.0	100.0	.000	0.00
Fuel Systems							
Throttle-body fuel injection	3.0	43	34.8	0.0	-34.8	-1.044	(16.15)
Multipoint fuel injection	5.0	91	58.0	100.0	42.0	2.100	41.56
Valve Train							
Overhead camshaft	3.0	111	43.7	97.1	53.4	1.602	64.52
4 valves per cylinder	5.0	141	36.3	97.1	60.8	3.040	93.04
Variable valve timing	6.0	142	0.0	97.1	97.1	5.826	149.20
Number of Cylinders [e]							
4-cylinder redesign	17.0	240	0.0	2.9	2.9	.493	7.55
6-cylinder redesign	17.0	442	0.0	0.0	0.0	.000	0.00
Transmission Technologies							
Torque converter lockup	3.0	50	45.5	66.7	21.2	.635	11.47
Electric transmission control	0.5	24	0.0	66.7	66.7	.333	17.35
4-speed automatic	4.5	225	16.8	0.0	-16.8	- .756	(40.96)
5-speed automatic	7.0	325	0.0	0.0	0.0	.000	0.00
Continuously variable transmission	8.0	325	0.0	66.7	66.7	5.333	234.89
5-speed manual	8.0	0	42.5	33.3	- 9.2	- .733	0.00
Rolling Resistance, Aerodynamics, and Weight							
Front-wheel drive	10.0	240	94.1	100.0	5.9	.590	15.35
Aerodynamics	2.3	40	18.3	100.0	81.7	1.879	35.43
Weight reduction, -10%	6.6	138	0.0	100.0	100.0	6.600	149.33
Electric power steering	1.0	45	0.0	100.0	100.0	1.000	48.78
Advanced tires, -10%	1.0	18	0.0	100.0	100.0	1.000	19.51
Advanced lubricants	0.5	2	0.0	100.0	100.0	.500	2.18

[a] Committee's adaptation of EEA estimates. See Appendix B.

[b] Weighted average cost based on 1990 distribution of engine sizes in the class before engine downsizing.

[c] Source: EEA, personal communication, October 2, 1991.

[d] Costs inflated to 1990 dollars by multiplying 1988 dollars by 1.084.

[e] Redesigned engines incorporate overhead camshaft, variable valve timing, and four valves per cylinder; the number of cylinders is reduced by two and the displacement is constant.

Table E-2 Illustrative Calculation of Fuel Economy Improvements and Incremental Costs for Subcompact Cars Using the Shopping Cart Method and Data in Table E-1

Technology	Market-Share Weighted Changes Effectiveness (mpg)	Market-Share Weighted Changes Cost ($)	Cost-Effectiveness (mpg/$)	Cumulative Retail Price Equivalent (1990 $)	Fuel Economy (mpg)
Base	---	---	---	0.00	30.46
Deceleration fuel restriction	0.132	0	--- [a]	0.00	30.59
Advanced lubricants	0.157	2.18	0.0720	2.18	30.75
Roller cam followers	0.445	12.46	0.0357	14.64	31.19
Engine redesign	0.155	7.55	0.0205	22.19	31.35
Friction reduction	0.551	28.80	0.0191	50.99	31.90
Aerodynamics	0.590	35.43	0.0167	86.41	32.49
Advanced tires	0.314	19.51	0.0161	105.93	32.80
Weight reduction	2.072	149.33	0.0139	255.26	34.87
Fuel system	0.332	25.41	0.0131	280.67	35.21
Variable valve timing	1.829	149.20	0.0123	429.87	37.03
Front-wheel drive	0.185	15.35	0.0121	445.22	37.22
Accessories	0.157	13.01	0.0121	458.23	37.38
4 valves per cylinder	0.955	93.04	0.0103	551.27	38.33
Valve system	0.503	64.52	0.0078	615.79	38.83
Transmissions	1.511	222.73	0.0069	838.52	40.35
Electric power steering	0.314	48.78	0.0064	887.30	40.66

[a] Since cost is zero, cost-effectiveness is very large.

NOTE: Technologies are listed in order of decreasing cost-effectiveness. Base fuel economy (31.4 mpg for MY 1990) is adjusted downward by 3 percent to account for safety and Tier I emissions standards. Calculations based on committee's adaptation of fuel economy, cost, and market-share data from EEA (1991a,b, and personal communication, October 2, 1991).

of Table E-2. The increments of fuel economy are added to the MY 1990 base value, which has been reduced by 3 percent to account for safety and Tier I emissions standards. To illustrate, a cumulative average fuel economy of 31.19 mpg corresponds to the incorporation of three technologies (deceleration fuel restriction, advanced lubricants, and roller cam followers) at a cumulative average cost of $14.64 per vehicle. The highest level of fuel economy achieved using this set of technologies is 40.7 mpg, achievable at a cumulative RPE of $887.30.

Figure E-1 shows the relationship of fuel economy to RPE for the subcompact car size class, taking account of the technologies appropriate to that size class from Table E-2. It is the same as "Case B" in Figure 7-4 of Chapter 7.

Tables E-3 and E-4 are similar to those described above, except that the fuel economy gains and associated costs are based on those estimated by SRI (1991). The MY 1990 market-share estimates are those of EEA (personal communication, October 2, 1991). Figure E-2 is the curve shown as "Case A" in Figure 7-4 of Chapter 7.

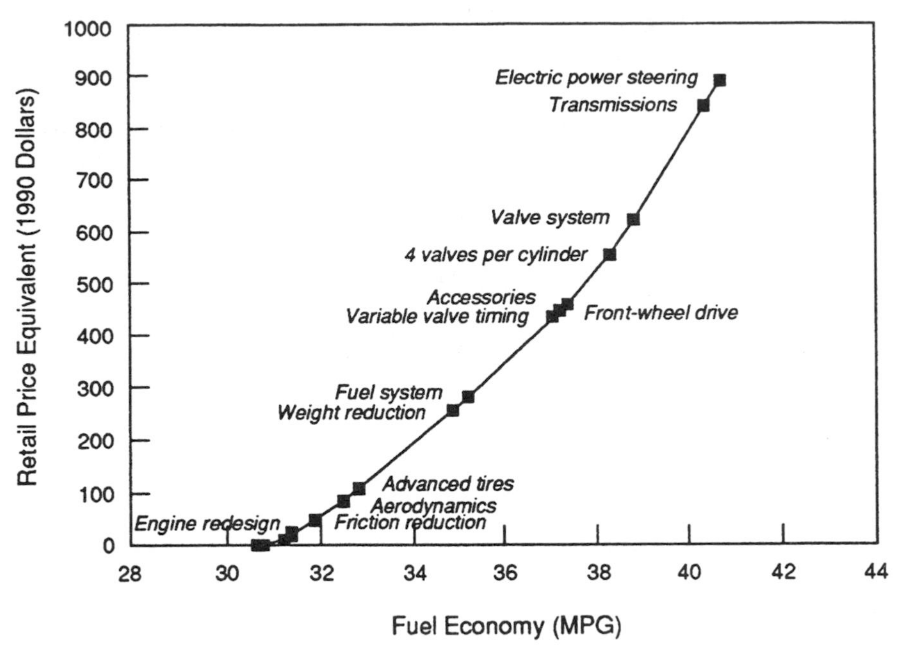

FIGURE E-1 Calculation of fuel economy costs for subcompact cars (data from Table E-2).

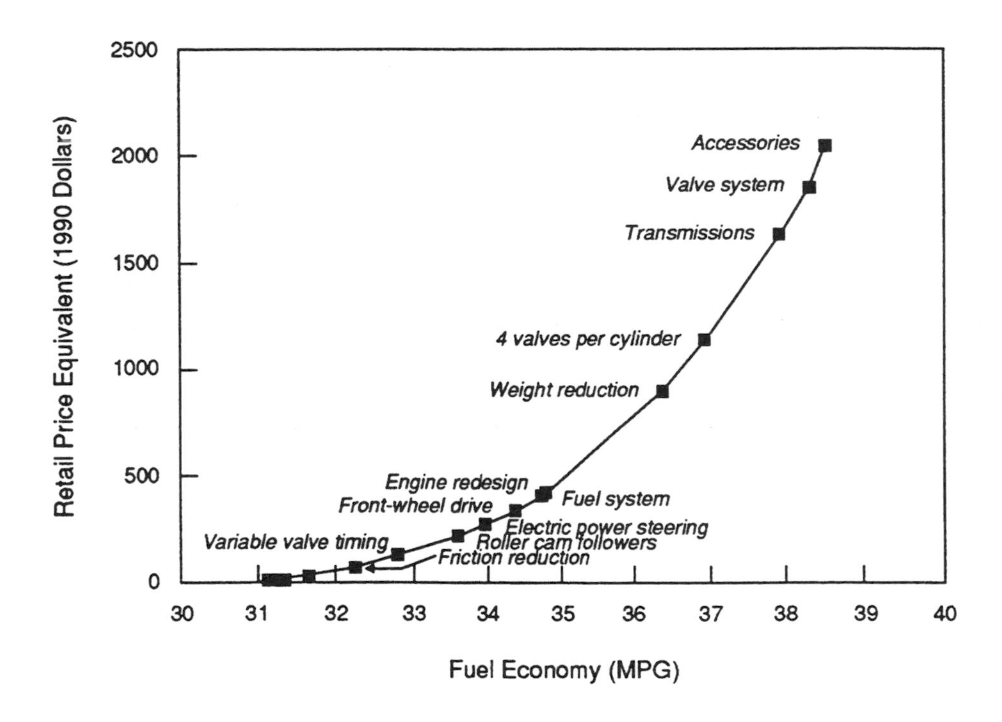

FIGURE E-2 Calculation of fuel economy costs for subcompact cars (data from Table E-4).

Table E-3 Technologies for Improving the Fuel Economy of Subcompact Cars (based on SRI data)

Technology	Fuel Economy Increase [a] (%)	Cost of Technology [a,b] (1990$)	Market Penetration (%) 1990 [c]	Market Penetration (%) Maximum	Change in Market Penetration (%)	Fuel Economy Increase x Change in Market Penetration (%)	Cost x Change in Market Penetration ($)
	(1)	(2)	(3)	(4)	(5)	(6)	(7)
Engine Technologies							
General							
Roller cam followers	1.7	65	29.2	100.0	70.8	1.168	48.02
Friction reduction, -10%	2.0	60	12.3	100.0	87.7	1.754	52.62
Accessory improvement	0.7	200	0.0	100.0	100.0	.700	200.00
Deceleration fuel restriction	1.0	5	58.0	100.0	42.0	.420	2.10
Compression ratio, +.5	2.0	1	0.0	100.0	100.0	2.000	1.00
Fuel Systems							
Throttle-body fuel injection	2.8	65	34.8	0.0	-34.8	- .905	(22.62)
Multipoint fuel injection	4.6	215	58.0	100.0	42.0	1.932	90.30
Valve Train							
Overhead camshaft	2.5	400	43.7	97.1	53.4	1.335	213.60
4 valves per cylinder	3.0	400	36.3	97.1	60.8	1.824	243.20
Variable valve timing	2.6	100	0.0	97.1	97.1	2.525	97.10
Number of Cylinders [d]							
4-cylinder redesign	8.1	600	0.0	2.9	2.9	.235	17.40
6-cylinder redesign	9.1	650	0.0	0.0	0.0	.000	0.00
Transmission Technologies							
Torque converter lockup	2.0	56	45.5	66.7	21.2	.423	11.85
Electric transmission control	0.5	122	0.0	66.7	66.7	.333	81.33
4-speed automatic	2.8	230	16.8	0.0	-16.8	- .470	(38.64)
5-speed automatic	3.3	530	0.0	0.0	0.0	.000	0.00
Continuously variable transmission	4.8	640	0.0	66.7	66.7	3.200	426.67
5-speed manual	4.8	0	42.5	33.3	- 9.2	- .440	0.00
Rolling Resistance, Aerodynamics, and Weight							
Front-wheel drive	0.5	26	94.1	100.0	5.9	.029	1.53
Aerodynamics	2.4	60	18.3	100.0	81.7	1.961	49.02
Weight reduction, -10%	5.0	470	0.0	100.0	100.0	5.000	61.00
Electric power steering	1.4	61	0.0	100.0	100.0	1.400	20.00
Advanced tires, -10%	1.0	20	0.0	100.0	100.0	1.000	3.00
Advanced lubricants	0.3	3	0.0	100.0	100.0	.300	2.18

[a] Committee's adaptation of SRI estimates. See Appendix B.

[b] Weighted average cost based on 1990 distribution of engine sizes in the class before engine downsizing.

[c] Source: EEA, personal communication, October 2, 1991.

[d] Redesigned engines incorporate overhead camshaft, variable valve timing, and four valves per cylinder; the number of cylinders is reduced by two and the displacement is constant.

Table E-4 Illustrative Calculation of Fuel Economy Improvements and Incremental Costs for Subcompact Cars Using the Shopping Cart Method and Data in Table E-3

Technology	Market-Share Weighted Changes Effectiveness (mpg)	Market-Share Weighted Changes Cost ($)	Cost-Effectiveness (mpg/$)	Cumulative Retail Price Equivalent (1990 $)	Fuel Economy (mpg)
Base	---	---	---	0.00	30.46
Compression ratio	0.628	1	0.628	1.00	31.09
Deceleration fuel restriction	0.132	2.1	0.0629	3.10	31.22
Advanced lubricants	0.094	3	0.0313	6.10	31.31
Advanced tires	0.314	20	0.0157	26.10	31.63
Aerodynamics	0.616	49.02	0.0126	75.12	32.24
Friction reduction	0.551	52.62	0.0105	127.74	32.79
Variable valve timing	0.793	97.1	0.0082	224.84	33.59
Roller cam followers	0.367	46.02	0.0080	270.86	33.95
Electric power steering	0.440	61	0.0072	331.86	34.39
Front-wheel drive	0.009	1.53	0.0059	333.39	34.40
Fuel system	0.323	67.68	0.0048	401.07	34.72
Engine redesign	0.074	17.4	0.0043	418.47	34.80
Weight reduction	1.570	470	0.0033	888.47	36.37
4 valves per cylinder	0.573	243.2	0.0024	1,131.87	36.94
Transmissions	0.957	481.21	0.0020	1,612.89	37.90
Valve system	0.419	213.60	0.0020	1,826.49	38.32
Accessories	0.220	200	0.0011	2,026.49	38.54

NOTE: Technologies are listed in order of decreasing cost-effectiveness. Base fuel economy (31.4 mpg for MY 1990) is adjusted downward by 3 percent to account for safety and Tier I emissions standards. Calculations based on committee's adaptation of fuel economy and cost data from SRI (1991) and market-share data from EEA, personal communication, October 2, 1991.

REFERENCES

Energy and Environmental Analysis (EEA), Inc. 1991a. Fuel economy technology benefits. Presented to the Technology Subgroup, Committee on Fuel Economy of Automobiles and Light Trucks, Detroit, Mich., July 31, 1991.

Energy and Environmental Analysis (EEA), Inc. 1991b. Documentation of Attributes of Technologies to Improve Automotive Fuel Economy. Prepared for Martin Marietta, Energy Systems, Oak Ridge, Tenn. Arlington, Va.

SRI International. 1991. *Potential for Improved Fuel Economy in Passenger Cars and Light Trucks*. Prepared for Motor Vehicle Manufacturers Association. Menlo Park, Calif.

APPENDIX F

COMMITTEE MEETINGS AND ACTIVITIES

1. Committee Meeting, May 13-15, 1991, Washington, D.C.

Welcome
Frank Press, President, National Academy of Sciences

Briefing Book on the United States Motor Vehicle Industry and Market (Version 1)
Robert Shelton, National Highway Traffic Safety Administration
John P. O'Donnell, Department of Transportation

Motor Vehicle Fuel Economy: A NHTSA Perspective
Jerry R. Curry, Administrator, National Highway Traffic Safety Administration

Rationale for Senate Bill S.279 and Expectations for the NAS Study
Senator Richard H. Bryan, Chairman, Subcommittee on Consumer, Committee on Commerce, Science and Transportation

The Automotive Industry: A Retrospective Look (1975-1991)
Deborah Gordon, Union of Concerned Scientists

Increasing Fuel Economy: How Far Can We Go?
Thomas H. Hanna, President and Chief Executive Officer, Motor Vehicle Manufacturers Association
Robert C. Stempel, Chairman and Chief Executive Officer, General Motors Corporation
Robert A. Lutz, President, Chrysler Corporation
Allan D. Gilmour, President, Automotive Group, Ford Motor Company

Gregory J. Dana, Vice President and Technical Director, Association of International Auto Manufacturers
Tokuta Inoue, Director, Toyota Motor Corporation and Director, Higashi-Fuji Research Center
E. Amito, Senior Vice President, American Honda Motor Company
Toni Harrington, Manager, Government/Industrial Relations, American Honda Motor Company
Karl-Heinz Ziwica, General Manager, Environmental Engineering, B.M.W. of North America, Inc.

Motor Vehicle Efficiency and Greenhouse Warming: Policy Implications
Rob Coppock, NAS Committee on Science, Engineering and Public Policy

Informing the Debate on Fuel Economy: The Needs of Congress
Congressman Philip R. Sharp, Chairman, Subcommittee on Power and Energy, Committee on Energy and Commerce

View of Labor on Improvements to Automotive Fuel Economy
Steve Beckman, International Economist, United Automobile Workers

Automotive Fuel Economy Studies at OTA: An Overview
Steve Plotkin, Office of Technology Assessment

Methodology Underlying EEA's Fuel Economy Projections: An Overview
K. G. Duleep, Director of Engineering, Energy and Environmental Analysis, Inc.

U.S. Safety Regulations for New Cars and Light Trucks
Donald Bischoff, Associate Administrator for Plans and Policy, National Highway Traffic Safety Administration
Barry Felrice, Associate Administrator for Rulemaking, National Highway Traffic Safety Administration

Automotive Fuel Economy & Safety
Stephen Oesch, General Counsel, Insurance Institute for Highway Safety
Clarence Ditlow, Executive Director, Center for Auto Safety

Future Emission Requirements
Karl Hellman, Environmental Protection Agency

2. **Workshop and Committee Meeting, July 8-13, 1991, Irvine, California**

AUTOMOTIVE TECHNOLOGY

Conventional Engines:
Charles Amann, Consultant
Erwin Nill, Mercedes-Benz Corporation
Leopold Mikulic, Mercedes-Benz Corporation
Masatami Takimoto, Toyota Motor Corporation

Advanced Engines:
Kim Schlunke, Orbital Engine Company
Karl-Heinz Neuman, Volkswagen of America

Drive Trains and Other Subsystems:

Variable Control Systems for Increasing Engine Efficiency Throughout the Full Power Range
Charles Mendler, Energy Conservation Coalition

The Impact of CFCs on Mobile Air Conditioning Systems
Kurt D. Hollasch, General Motors Corporation

Fuels and Lubricants:

Lubricants and Engine Friction Reduction
David Hoult, Massachusetts Institute of Technology

Trends in Fuel Composition and Impact on Fuel Economy and Emissions
Joe Colucci, General Motors Corporation

Materials Considerations in Vehicle Design and Operation:
David Parker, Aluminum Association
David Schlendorf, ALCOA
Ronald McClure, ALCOA
Alan Seeds, Alcan Aluminum Corporation
Randy Suess, Dow Chemical Company
Peter Peterson, U.S. Steel Corporation

CONCEPT CARS AND PROTOTYPES

Advanced Light Vehicle Concepts
Amory Lovins, Rocky Mountain Institute

Prototypes and "Best in the World Cars": Overview and Lessons Learned
Deborah Bleviss, International Institute for Energy Conservation

THE CAR AS A SYSTEM: FUEL ECONOMY POTENTIAL AND PROSPECTS

Allan D. Gilmour, Ford Motor Company
Katsumi Suzuki, Toyota Motor Corporation
Masatami Takimoto, Toyota Motor Corporation
Ronald R. Boltz, Chrysler Corporation
Takefumi Hosaka, Honda R&D Co., Ltd., Tochigi Center
Ronald H. Haas, General Motors Corporation
Yoichiro Kaneuchi, Nissan Motor Company, Ltd., of Japan
Yoshiaki Danno, Mitsubishi Motors Corporation

Economic Effects of Tightening CAFE Standards
Michael J. Boskin, Chairman, Council of Economic Advisers, (by Video TeleConference)

FUEL ECONOMY AND COST PROJECTION METHODOLOGIES

An Engineering Assessment of Fuel Economy Opportunities
Marc Ross, University of Michigan

A System for Estimating Fuel Economy Potential Due to Technology Improvements
Richard Andrews, University of Michigan
James Berger, Purdue University
Murray Smith, University of Canterbury

Fuel Economy Projections to Year 2010
K. G. Duleep, Energy and Environmental Analysis, Inc.

Technology Improvement Incremental Cost Analysis
Henry Allessio, Easton Consultants, Inc.

SAFETY

Safety vs. Fuel Economy: A Trade-off
B. J. Campbell, University of North Carolina

Vehicle Downsizing versus Vehicle Downweighting: Implications for Safety
Charles Kahane and Terry Klein, National Highway Traffic Safety Administration

Potential Improvements in Occupant Packaging to Offset Vehicle Weight Reduction
Donald Friedman, Liability Research, Inc.

An Econometric Model of Fuel Economy and Passenger Vehicle Highway Fatalities
J. Daniel Khazzoom, San Jose State University

How to Save Fuel and Reduce Injuries in Automobiles
Leon Robertson, Nanlee, Inc.

Public Policy and the Automotive Safety Issue
Benjamin Kelley, Institute for Injury Reduction
(presented by Craig McClellan, McClellan & Associates)

A Rational Approach to the Safety Debate
Phil Haseltine, American Coalition for Traffic Safety

CLEAN AIR ACT AMENDMENTS: WHAT EFFECT ON FUEL ECONOMY?

Automotive Industry Perspective:
Kelly Brown, Ford Motor Company
Gregory Dana, Association of International Automobile Manufacturers
Richard Penna, Toyota Motor Corporation

ECONOMIC AND REGULATORY CONSIDERATIONS IN IMPROVING FUEL ECONOMY

Review of Regulatory Options at the Federal and State Levels
Ralph Cavanagh, Natural Resources Defense Council

Regulations and Market Forces
Gary R. Fauth, Charles River Associates

Fuel Economy: A Customer's Viewpoint
Robert Leone, Boston University

Cost-Effectiveness of Increased Fuel Economy
John DeCicco, American Council for an Energy-Efficient Economy

CONSUMER BEHAVIOR AND DEMAND FOR CARS

Behavioral Studies and Consumer Demand
Willet Kempton, Princeton University

Consumer Behavior and New Car Purchase
Steve Barnett, Nissan North America

Survey Studies and Consumer Demand
Dave Power, J. D. Power, Inc.

CAFE and Consumer Behavior
George Borst, Toyota Motors Sales U.S.A., Inc.

BARRIERS TO INTRODUCTION OF HIGH FUEL ECONOMY VEHICLES IN THE U. S. MARKET

Automotive Industry Perspective
Ken Kohrs, Ford Motor Company

Potential for Improving Fuel Economy of Passenger Cars and Light Trucks
Norman Stoller, SRI International

Should Consumer Preferences for Comfort, Safety and Performance in a Low Energy Cost World be Considered a Barrier?
Fred Smith, Competitive Enterprise Institute

Resources, Motivation, and Lead Time
Tom Feaheny, Consultant

A Concept to Improve the Fuel Economy of the Nation's Motor Vehicles
Patrick Raher, Mercedes-Benz Corporation

LIGHT TRUCK AND VAN POLICY

Basis for Current Regulations on Fuel Economy and Safety of Light Trucks and Vans
Orron Kee, National Highway Traffic Safety Administration

Current Purchase and Use Patterns of Light Trucks and Vans
William Bostic, U.S. Department of Commerce

Unique Fuel Economy Considerations for Light Trucks and Vans vis-a-vis Passenger Cars
James Englehart, Ford Motor Company
Yoichiro Kaneuchi, Nissan Motor Company, Ltd., of Japan

LATE PAPER

Parallels Between U.S. and Australian Automotive Fuel Economy Problems
Peter Anyon, Australian Federal Government

WRAP UP
Thomas H. Hanna, Motor Vehicle Manufacturers Association
Gregory J. Dana, Association of International Auto Manufacturers
Ralph Cavanagh, National Resources Defense Council

3. Technology Subgroup Meeting, July 31, 1991, Detroit, Michigan

Norman Stoller, SRI International
Phil Amos, SRI International
Larry K. Ranek, SRI International
Pamela J. Olson, SRI International
Marcel Halberstat, Motor Vehicle Manufacturers Association
Thomas H. Hanna, Motor Vehicle Manufacturers Association
K. G. Duleep, Director of Engineering, Energy and Environmental Analysis, Inc.

4. Safety Subgroup Meeting, August 21-22, 1991, Washington, D.C.

Leonard Evans, General Motors Corporation
Ernest Grush, Ford Motor Company
Brian O'Neil, Insurance Institute for Highway Safety
B. J. Campbell, University of North Carolina
Clarence Ditlow, Center for Auto Safety
Mark Edwards, National Highway Traffic Safety Administration
Charles Kahane, National Highway Traffic Safety Administration
Terry Klein, National Highway Traffic Safety Administration
Robert Shelton, National Highway Traffic Safety Administration

5. Committee Meeting, August 23-25, 1991, Cambridge, Massachusetts

No presentations were made at this meeting.

6. Technology Subgroup Meeting, September 5-6, 1991, Detroit Michigan

Chrysler Corporation
Robert Lutz
Gordon Allardyce
Beverly Bunting
Van Bussmann
Francois Castaing
Arnold DeJong
Thomas Gage
Peter Gilezan
James Rickert
Richard Schaum
Robert Sexton
Al Slechter

Ford Motor Company

Allan Gilmour
Dan Ahrns
Chris Aliapoulis
Bob Bacigalupi
Dick Baker
Peter Beardmore
Chinu Bhavsar
Kelly Brown
Jim Endress
Haren Gandhi
Ed Hagenlocker
Bob Himes
Mike Jordan
Thomas Kenney
Ken Kohrs
David Kulp
John LaFond
Pete Pestillo
Helen Petrauskas
Jeff Pharris
Norm Postma
Bill Quinlan
Bob Rankin
Bob Roethler
Al Simko

General Motors Corporation

Robert Stempel
Jack Armstrong
Lewis Dale
Harry Foster
Nicholas Gallopoulos
Ronald Haas
Livonia Plant
Donald Runkle
Leon Skudlarek
Thomas Stephens
Gerald Stofflet
Tom Young

Honda

E. Amito
Toni Harrington
Takefumi Hosaka
H. Kano
Atsushi Totsune

7. **Impacts Subgroup Meeting, September 16, 1991, Washington, D.C.**

Charles River Associates
David Montgomery

Chrysler Corporation
Van Bussmann
Tom Gage
Al Slechter

Ford Motor Company
Allan D. Gilmour
Kelly Brown
Bobbi Koehler-Gaunt
Michael Jordan
Peter Pestillo
Helen Petrauskas
Susan Shackson
Greg Smith
Martin Zimmerman

General Motors Corporation
Lewis Dale
Michael DiGiovanni
George Eads
Harry Foster
Stephen O'Toole
Gerald Stofflet

8. **Technology Subgroup Meeting, September 18, 1991, Washington, D.C.**

Conventional and Advanced Automotive Fuel Economy Technology: Future Potential and Prospects
Gary Rogers, FEV, Inc.

Post 2001 Technology Options: Power Trains, Aerodynamics, Electric Vehicles, Hybrids and CAFE Alternatives
Paul McCready, Aerovironment, Inc.

Roundtable discussion on Conventional Technology, Advanced Technology and CAFE Standards
John DeCicco, American Council for an Energy-Efficient Economy
Charles Mendler, Energy Conservation Coalition
Marika Tatsutami, Natural Resources Defense Council

9. Committee Meeting, September 19-21, 1991, Washington, D.C.

Raymond Wassel, Board on Environmental Studies and Toxicology,
National Research Council

10. Committee Meeting, October 14-16, 1991, Washington, D.C.

No presentations were made at this meeting.

11. Meeting of the Subgroups on Emissions and Environment, October 24, 1991, Washington, D.C.

Toyota Motor Corporation
Saburo Inui
Tadao Mitsuta
Ryuzo Oshita
Richard Penna

Mitsubishi
Yoshiaki Dann
Steve Sinkez

General Motors
Jack Benson
Lewis Dale
Samuel Leonard
Stephen O'Toole
Gerald Stofflet
Richard Taylor
Robert Wiltse

Ford Motor Company
Kelly Brown
Richard Baker
Haren Gandhi
Helen Petrauskas

Volkswagen of America
Leonard Kata
Karl Heinz-Neumann

Mercedes-Benz Corporation
Klaus Drexl
William Kurtz
Patrick Raher

Environmental Protection Agency
Karl Helman

12. Meeting of the Subgroup on Standards and Regulations, November 4, 1991, Washington, D.C.

Toyota Motor Corporation
Charles Ing
Saburo Inui
Tetsushi Itoh
Richard Penna
Kazuko Sherman
Junzo Shimizu
Katsumi Suzuki

General Motors Corporation
George Eads
William Ball
Harry Foster
James Johnston
Gerald Stofflet

Honda
E. Amito
Toni Harrington

Ford Motor Company
Kelly Brown
Allan Gilmour
Susan Sheckson
Martin Zimmerman

Chrysler Corporation
Ronald Boltz
Van Bussmann
Thomas Gage
Robert Liberatore

Natural Resources Defense Council
Ralph Cavanagh

13. Committee Meeting, November 11-13, 1991, Washington, D.C.

No presentations were made at this meeting.

14. Technology Subgroup Meeting, November 22, 1991, Washington, D.C.

No presentations were made at this meeting.

15. Committee Meeting, December 12-14, 1991, Washington, D.C.

No presentations were made at this meeting.

APPENDIX G

BIOGRAPHICAL SKETCHES OF COMMITTEE MEMBERS

Committee on Fuel Economy of Automobiles and Light Trucks
Energy Engineering Board
National Research Council

Richard A. Meserve (chairman) is a partner in the law firm Covington & Burling of Washington, D.C. His educational background includes a J.D. from Harvard University and a Ph.D. in applied physics from Stanford University. He served as legal counsel to the President's Science Adviser for the period 1977-1981. He is now chairman of the National Research Council's (NRC) Panel on Cooperation with the USSR on Reactor Safety and previously chaired the Committee to Provide Interim Oversight of the Department of Energy's Nuclear Weapons Complex. He is a member of the NRC Committee on Scientific Responsibility and the Conduct of Science.

Gary L. Casey is former director, Advanced Technology, at Allied-Signal, Inc., Troy, Michigan, and has managed a variety of R&D functions involving brake, suspension, and engine control systems. He has also served as director of engineering at Mercury Marine, which manufactures marine propulsion systems. He is a mechanical engineer by training, has over 20 years of experience in automotive R&D, and is an adjunct professor at Wayne State University.

W. Robert Epperly is president of Epperly Associates, Inc., a consulting firm in New Canaan, Connecticut. He was previously chief executive officer of Fuel Tech N.V., a company engaged in development and commercialization of combustion technology to improve efficiency and reduce emissions. Earlier, he was at Exxon Research and Engineering Company, where he ended 29 years of service as general manager, Corporate Research. He served on the NRC's Committee on Synthetic Fuels Facilities Safety and chaired its Committee on Cooperative Fossil Energy Research. He holds an M.S. in chemical engineering from Virginia Polytechnic Institute.

Theodore H. Geballe is a professor of applied physics and material sciences at the Department of Applied Physics, Stanford University. Past service at Stanford includes chairman, Department of Applied Physics, and chairman, Center for Materials Research. Previously, he served as head, Department of Low Temperature and Solid State Physics in the Physical Research Laboratory, Bell Telephone Laboratories, Murray Hill, New Jersey. He is a member of the National Academy of Sciences, the American Academy of Arts, and a fellow of the American Physical Society.

David L. Greene is a senior research staff member at Oak Ridge National Laboratory, Tennessee. His work has focused on national policy issues related to transportation energy use, efficiency, and alternative fuels. He is chairman of the section on Environmental Concerns of the NRC's Transportation Research Board and recent chairman of the Committee on Conservation and Transportation Demand. He has a Ph.D. from Johns Hopkins University.

John H. Johnson is presidential professor and chairman, Department of Mechanical Engineering and Engineering Mechanics at Michigan Technological University, Houghton. His research work includes combustion studies, hybrid engines, tribology, emissions, and air pollution. He has served on committees of the National Academy of Sciences, Office of Technology Assessment of the U.S. Congress, and National Aeronautics and Space Administration. He holds a Ph.D. in mechanical engineering from the University of Wisconsin.

Maryann Keller is managing director and automotive analyst with the brokerage firm of Furman Selz Incorporated, New York. Her work for the past 20 years has focused on the automotive industry. Her previous positions were with the investment advisory firms of Vilas-Fischer Associates, Inc., Paine Webber Mitchell Hutchins, and Kidder, Peabody & Company, Inc. She was a participant in the Massachusetts Institute of Technology's four-year study of the automotive industry, currently serves on the Committee to Assess Advanced Vehicle and Highway Technologies of the NRC's Transportation Research Board, and is president of the Society of Automotive Analysts. She holds an M.B.A. from the City University of New York.

Charles D. Kolstad is associate professor, Institute for Environmental Studies and Department of Economics, University of Illinois, Urbana, and a member of the NRC's Energy Engineering Board. He has also been on the faculty of the Massachusetts Institute of Technology and the staff of the Los Alamos National Laboratory. For over 15 years he has been involved in research on energy and environmental economics and is the author of over 80 scholarly articles, books, chapters, and reports. He holds a Ph.D. from Stanford University.

Leroy H. Lindgren is vice president, Manufacturing Planning Systems, Rath & Strong, Inc., Lexington, Massachusetts, a consulting firm that specializes in manufacturing operations, production planning, facilities design, and costing. He also served there as director of technical services and vice president of policy and planning and has extensive experience with the U.S. automotive industry. He was a member of the National Academy of Sciences' Committee on Motor Vehicles and served as a consultant to the Department of Transportation, Department of Energy, and the

Environmental Protection Agency. He holds a B.S. in mechanical engineering from the Illinois Institute of Technology and has served as adjunct associate professor at Boston University.

G. Murray Mackay is head of the Accident Research Unit, Automotive Engineering Center, University of Birmingham, England, where he has been a reader in traffic safety. His research interests include vehicle design and collision performance, epidemiology of transport accidents, traffic engineering, and the biomechanics of injury. He is fellow of the Institution of Mechanical Engineers and has served as director and president of the American Association for Automotive Medicine. He holds a Ph.D. and D.Sc. from the University of Birmingham.

M. Eugene Merchant is senior consultant at the Institute for Advanced Manufacturing Sciences, Cincinnati, Ohio. Previously, he was director, Advanced Manufacturing Research at Metcut Research Associates, Inc., and principal scientist for manufacturing research at Cincinnati Milacron Inc. He is a member of the National Academy of Engineering and has served on the NRC's National Materials Advisory Board and Manufacturing Studies Board. He is past president of the Society of Manufacturing Engineers, the International Institution for Production Engineering Research, American Society of Lubrication Engineers, and the Federation of Materials Societies. He holds a D.Sc. from the University of Cincinnati, where he has been an adjunct professor of Mechanical Engineering.

David L. Morrison is technical director, Energy, Resource and Environmental Systems Division, The MITRE Corporation, McLean, Virginia. He was previously president of the IIT Research Institute and director of Program Development and Management, Battelle Memorial Institute. He is a member of the NRC's Energy Engineering Board, has served on the NRC's National Materials Advisory Board, and most recently was chairman of the Committee on Alternative Energy R&D Strategies, whose work resulted in the publication *Confronting Climate Change: Strategies for Energy, Research, and Development* (1990). He holds a Ph.D. in chemistry from the Carnegie Institute of Technology.

Phillip S. Myers is emeritus distinguished research professor, and former chairman, Department Mechanical Engineering, University of Wisconsin, Madison. He is a member of the National Academy of Engineering and fellow of the American Society of Mechanical Engineers and was the 1969 National President of the Society of Automotive Engineers. He was a member of the NRC's Committee on Production Technologies for Liquid Transportation Fuels, whose work resulted in the publication *Fuels to Drive Our Future* (1990). His research interests are in internal combustion engines, combustion processes, and fuels. He holds a Ph.D. from the University of Wisconsin.

Daniel Roos is professor of Civil Engineering, and director of the Center for Technology, Policy, and Industrial Development, Massachusetts Institute of Technology, Cambridge. He is also the director of the International Motor Vehicle Program at MIT, whose reports include *The Machine That Changed the World* (1990) and *The Future of the Automobile* (1986). He has also been director of MIT's Center for

Transportation Studies. He has served as chairman of the Paratransit Committee of the NRC's Transportation Research Board and is chairman of the Committee to Assess Advanced Vehicle and Highway Technologies. He holds a Ph.D. in Civil Engineering from MIT.

Patricia F. Waller is director of the University of Michigan Transportation Research Institute, Ann Arbor. Previously, she was research professor, School of Public Health, University of North Carolina, and director of the university's Injury Prevention Research Center. She also served as associate director for driver studies of the university's Highway Safety Research Center. She chairs the NRC's Transportation Research Board Council on Intergroup Resources and is a member of their Research Technology and Coordinating Committee for the Federal Highway Administration and committees on Planning and Administration of Transportation Safety; Motor Vehicle Size and Weight; and Alcohol, Other Drugs and Transportation. She is a psychologist and holds a Ph.D. from the University of North Carolina.

Joseph D. Walter is director, Central Research, at Bridgestone-Firestone, Inc., Akron, Ohio. He is an expert in polymers and composites and has 20 years of experience in tire design and rolling friction. He is editor and/or author of book and articles on the mechanics of pneumatic tires and has done advanced design work on composite wheels for automobiles. He is a member of the Accreditation Board for Engineering and Technology. He holds a Ph.D. in engineering from Virginia Polytechnic Institute and an M.B.A. in finance from the University of Akron.

INDEX

INDEX